"ABSOLUTE TRUTHS", OR UNCENSORED FREE SPEECH, ONE OF THEM LEADS US TO HELL

It would be impossibe to investigate our Universe
without a set of principes and a scientifc logic

H O S S E I N   D A L L A L B A S H I

TABLE OF CONTENTS

A Comparative Analysis of Lenin's and Engels's Theory of Knowledge, (epistemology) & a Critique of Lenin's Theory of Knowledge.

I wrote this research paper as a term paper assignment for a philosophy class, at Los Angeles Cal State University in 1979. I was both majoring in philosophy and economics.

A NOTE FROM THE AUTHOR, AND A DEDICATION

IN ALL MY COLLEGE, AND university years, dating back to early sixties, up to early eighties, while I was going through school regular academic courses, I was intellectually fascinated, consumed, and preoccupied with the concept of dialectical materialist logic. I always thought that it would be impossible to attempt to pursue the study of any of the natural sciences, or social sciences, without a scientific outlook or a scientific logic, within the context of which, one could analyze, and determine the degree of scientific nature, relevancy and validity of the subject matter undertaken. Without having and consciously using a philosophically scientific outlook, as a guide, to seek scientific knowledge, would be like walking in complete darkness, not being able to have any sense of directions, east, west, south, north, no concept of time, location, and more than anything else, who you are in relationship to your surroundings. I leave it to your imagination, as to how far you could go, accomplishing anything, in that environment. As a young man, full of unbounded energy, enthusiasm, dreams, and aspiration to help the positive social forces, to bring about a more humanized society, I started my studies with social sciences, political science, economics, and philosophy. It was my acquaintance with dialectical materialist logic that made me realize that without comprehensive studies of a scientific philosophy, involving extensive studies, and understanding of philosophical and scientific concepts of matter, motion, space, time, causality and also natural sciences, such as atomic physics, chemistry, biology and genetics, and all the other sciences derived from them, it would be impossible to understand any sciences whatsoever, in a profound fashion. Awareness and knowledge of Dialectical materialist logic would make us conscious of our intellectual shortcomings and ignorance of our Universe, and how its unlimited parts and aspects work, **because, for me, science means the ability to unlock the ever - changing secrets of the Universe.** And if our tools of scientific investigation are faulty, and antiquated, then our end results would also be naïve and distorted. **In contrast, without awareness, and knowledge of dialectical materialist logic, we would be superficially memorizing cliché statements, useless formulas, and engage in meaningless generalities, without profoundly understanding the nature of our Universe,** which is the basis of any science we study. We would be studying different sciences, as if they had nothing to do with one another; mindlessly, and unconsciously using **"formal logic, and Aristotelian logic" to constantly legitimize the separation of different parts of nature and also natural sciences,** and unfortunately, all sciences into unrelated issues; and for that matter, Everything in nature whereas, dialectical materialist logic would put **all sciences within the context of one integrated unit,** and show the individuality and interrelationship of all sciences to one another. Each science would have to defend its existence, validity, and relevance within the context of that integrated unit, or it can't stand alone for too long. There has to be a unified approach in science, **which unfortunately doesn't exist in the educational system of the United States.** Every time, you get confused on this issue, put our solar system in your imagination, and ask yourself: *which* part of *it relates* to regular physics, or atomic physics, which part relates to chemistry, which part to biology, which part to genetics, and also to unlimited offspring natural sciences derived from natural sciences. **In nature, all natural sciences come together, and are integrated**. But, it is human beings who, out of ignorance and short- sighted convenience, or lack of deep scientific understanding, have created artificial separation of the Universe, into unrelated parts, and in their limited minds, attempt to study the parts independently of one another. This can only lead to severe, negative educational, cultural and practical consequences of all kinds for all mankind to bear, because of this narrow- minded approach to life. When that approach is used in social sciences, economics, sociology, political science, history, philosophy, even psychology, the result is a disaster, a nuclear bomb devastation. **Instead of studying society as a social organism, very much like a human body, we study different aspects of the society in separation from one another.** We study social institutions, without realizing why these institutions came into being, responding to certain individual, and social needs, and formation of an economic system. That Is why studies of social sciences, in my opinion, in U.S. do not make sense, and they are very sterile, and shallow. A non- politicized dialectical materialist logic makes us aware of this intellectual, and cultural narrow-mindedness, and offers an alternative unified approach to studying our Universe, and societies: and this is clearly the right step *in a right direction*. ***But, I have made an observation which provides some clarity on two different problems, associated with the*** **concept of dialectical materialist logic,** in order to defend itself as a scientific logic: First, at the hand of the Left politics and ideology, dialectical materialist logic has become too politicized, and therefore almost useless, as a scientific logic; and second, the Right politics and ideology have developed an absolutely uncompromising hatred for dialectical materialist logic, because Karl Marx, and Frederick Engels, the founders of Marxist Socialism had used dialectical materialist logic to analyze the capitalist system and conclude that it would have to be replaced by a more humanized system. This has been considered an unpardonable sin by the entire capitalist educational system. Marxist Socialism won itself this uncompromising hatred and lack of lifetime pardon, and acceptance by the entire capitalist educational system, which is so deep and profound, that they have absolutely trashed out the significance of dialectical materialist logic as a genuine scientific logic, which is, in my opinion, a thousand times more preferable, and useful, as a scientific outlook, than **formal logic, and Aristotelian logic**, which the capitalist system teaches us on college and university levels. Mathematics is used by many legitimate, and corrupt professions for multiple uses, and reasons. An atomic physicist uses it to arrive at the formula of a nuclear atomic bomb, which would destroy life in general beyond our imagination. Unscrupulous Wall Street bankers use mathematics to rip off unsuspecting ordinary people by billions of dollars, depriving them of their life time savings. A thief uses it to decipher the combination lock of a bank safe to steal money. A pharmacist uses mathematics to correctly combine various atomic elements to make certain medications, that would save the lives of fatally ill people. A house wife uses it to make sure that she has not gone beyond her budget and priorities of the month.

A perfectly legitimate question arises in our mind, since there are so many abuses of Mathematics, with harmful effects in the entire society: should we outlaw mathematics because of its use, and abuse by so many different undesirable people with immoral and unethical intentions? We would be short changing ourselves if we did, and few people would be in favor of this. The same thing should apply to the concept of any logic, and in this case the logic of dialectical materialism. If dialectical materialist logic were to be developed further scientifically by our natural scientists globally, as it failed to do so for almost seventy years, in the Soviet Union, because Leninism contained its natural development, and Soviet scientists were forced to interpret dialectical materialist logic from a narrow approach of being at the service of Leninism, and the Soviet Communist Party ideology, it could first create fundamental transformation in all nat- ural sciences, beyond our belief and expectation; and then would create a fundamental scientific renaissance in every aspect of society, which would make our existing scientific system, more of a joke. Dialectical materialist logic is an intellectual scientific achievement, belonging to all mankind, and it is a profound error to dismiss it as useless, harmful and attempt to eliminate it altogether from our educational system. Our educational system could conveniently ignore, as it has, *Marxist economic analysis, which involves a critique* of the capitalist system, but, the case for dialectical materialist logic, as an alternative form of logic is different and should be included in college, and university curriculums. If our intentions are to create intelligent individuals and citizenry in our society, which I doubt very much, then dialectical materialist logic is a useful tool for that rea- son. I am absolutely sure that by teaching dialectical materialist logic in our schools, colleges, and universities, as an alternative logic, along with formal logic, and Aristotelian logic, which is the only logic taught in our academic institutions, we would allow the minds of our students expand, becoming much more alert inquisitive smarter, shrewder, profounder, more thoughtful, and more intelligent, with better quality education. Dialectical materialist logic is like mathematics. It is universal and is not necessarily associated with any economic systems. People with all political orientations use mathematics. We could safely arrive at the conclusion that mathematics is not political, and ideological. Education should not be a system of manipulations, designed to prove or disprove certain ideologies. **It should be an honest attempt to decipher the secrets of the Universe, and soci- ety, and be at the service of mankind.** But, unfortunately, so far, in the kinds of governments we have had, all sciences have become subservient to certain politics, and ideologies. The thieves and the spiritual could use it for different intentions, but it does not belong to either one of them. As a matter of fact, dialectical materialist logic would even make a theologian much wiser, and smarter in relating to his religious followers. I think one could see the evidence of that in **Pope Francis, because as the story goes, and his patterns of reasoning, and positions, associated with various global issues, confirm, he has a degree in chemistry**. It would make a difference if a theologian is acquainted with natural sciences. Maybe, it should be a requirement for all pastors to study natural sciences, before they do sermons at churches on Sundays. After all, studies of natural sciences would enable us to become better acquainted with the Planet Earth, one of the master pieces of the Good Lord's creations, and that is where genuine enlightenment would begin. **Engels believed, dialectical materialism is no longer a philosophy, above all sciences, it is the embodiment of all natural sciences, and modern chemistry, based upon this definition, is definitely one of the main pillars of dialectical materialist** logic. **I think very few natural science students, in this country, are acquainted with it.**

They honestly don't know what they are missing. Both, mathematics, and dialectical materialist logic are intellectual achievements of all mankind. **No society, group, race, or culture has the exclusivity or monopoly to any intellectual achievements. They belong to all of us, because, they originate from the collective intellect, and wisdom of all mankind, and have universal practical uses, beneficial to all.** As for my work en- titled, "self motioned atomic dialectical materialist logic", which I wrote it in 1979, and since then, I have gone way beyond that. This is the nature of dialectical materialist logic. If you keep studying it, there is no limit to its horizon, and advancement. **We could never reach a final point at which we would say: "now I get it". The limit is the Universe itself.** It would definitely take you to places, where you never intended to go. If you read another book, that I wrote, called: 1Formal logic, Aristotelian logic vs. evolutionary integrated atomic logic", you could see my intellectual growth and advancement in the understanding of the highly volatile dialectical materialist logic. You would understand why in the Universe, an atomic particle is in one place in any given time, and literally in trillions of others, at the same time. Whereas, formal logic or Aristotelian logic teach us that every atom, every object, every f lower, every animal, every human being has well defined absolute distinction, absolute existence and absolute place in the Universe, in any given time and space, *unrelated to anything else in the Universe*. *In contrast, dialectical materialist logic teaches us that the Universe is atomically integrated*, and the study of any part of the Universe would necessarily involve unlimited number of other issues, even though, it may not be convenient for us to go beyond a narrow, and short sighted beginning step, **which studies everything in isolation from other things**, being usually the case, in our educational system. For that reason, **I would like to dedicate this short humble research book, on dialectical materialist logic, to all men and women of scientific philosophy and natural sciences**, globally, specially atomic physics, chemistry, biology, and genetics, and unlimited offspring sciences derived from them. Together, they form the basis of any human civil society. Now, you know why it is hard to swallow that formal logic, Aristotelian logic should dominate *our entire global cultures, as the only form of reasoning, and rationality;* while through experience, and human intuition, we know there are too many things wrong, but we are not able to pinpoint them. Why?, because, the bicycle of knowledge and wisdom that we are riding does not allow us to go through the oceans, much less provide us with the opportunity to travel to the horizons of other Planets, or even to other solar systems, and the entire Universe. For that, we need an atomically crafted spacecraft, while we could use the bicycle to go to the neighborhood liquor store to get a six pack of Mexican beer. Each is useful in its own time and place. The bicycle that limits our ability to travel to unknown places, and stunts our intellectual development, and we are not aware of it, **is the formal logic Aristotelian logic, that we teach in our academic institutions.**

No wonder, every four years, a complete idiot from both the Democratic and Republican parties emerge, and with a new dance and song, and a lolli pop wants " to make us become a great country again", and we can't make the connections, until his term of office is over, four years later. That happens, because we are not intellectually trained to analyze things in historical perspective, and interrelationship in advance, before we actually go through them. We are about to finish eight years of a well spoken Con artist, Barak Hussein Obama, and once again, indulge in getting either Donald Trump, or Hillary Clinton. Maybe, that is what we deserve.

Scientists know in advance that certain poisonous substances **will** kill, but the regular folks, who do not believe in science, take the poison to find out if it is indeed true. Obviously, they won't live through the ordeal to see the consequences. But, their family survivors realize that the deceased should have taken science more seriously. But now, it is too late. A good example of this would be when Conservative Republican talk shows, in the United States, constantly ridicule, and mock the ideas of climate change, by man's unwise economic, and political intervention in nature, believing that the rise of the oceans, in the coastal cities, that creates untold emotional suffering and devastating economic and financial damages, and destructions, and the drying of some of the lakes and rivers around the world, have always been around and will continue to exist, regardless of what human beings would do to diminish their occurrences. Natural scientists tell us a different story. But, in this country, we could not care less what sciences would tell us, we would go right ahead until one day, all mankind would pay a much heavier price for this very obvious stupidity. But, who cares, because the generation that ignored the warnings of nature, will not be around, any more, to suffer the consequences of their actions. They poison the impressionable minds of our children, to accept whatever injustices we do to nature, the basis of our survival, hiding behind the protection of our Constitutional rights of free speech. These talk show people do not bother inviting some pro, and con scientists, to explain the issues, from a genuine scientific approach. They present, and interpret science, within a few very broad cliché statements they have memorized, repeating them on and on every single day, until, they themselves would believe it. Listen to 870, AM "The Answer", radio station, in Los Angeles, and you would have what I am telling you, confirmed, over, and over. I consider this a very profound abuse of Constitutionally guaranteed freedom of speech, which is, more precisely stated, brain- washing the people, as opposed to enlightening them. Science should not have anything to do with any kind of political, and ideological orientations of any kind, but unfortunately, we are unconsciously trained to believe, that everything should be seen from the point of views of politics, ideology, with religion in the background. We shift the consequence of this type of ignorance to the newer generations. We constantly go through unnecessary trials and errors, in every aspect of life, for establishing the truths of things, instead of using a scientific outlook, and scientific logic as a guideline, to verify the truths, in the beginning, and not in the end. Winston Churchill, the British Prime Minister, once made a remark that the: "Americans always make the right decisions, but after they have tried, gone through everything, every possible error". The major problem is that, so far, we have been at- tempting to understand everything, specially nature, from the positions of ideology, an politics; and this has been the unpardonable sin of the Right, and the Left. Now, it is time to de-ideologize, and de politicize our habit of investigation to a scientific approach. If you don't believe it, try to build a highway bridge, or go through an open heart surgery, based upon ideology and politics, and let us see what the results would be. I bet you that you will not be too anxious, or be the first person, to drive over that bridge, and Just, pray to God that you are not the patient, going through that open heart operation ideologically. But, we use ideology and politics to recommend an open heart operation, and the building of highway bridges for others. As long as somebody else will pay for the price, we couldn't care less. AT this writing, an oil pipe line is being built, coming from Canada, going through North Dakota, through the indigenous land, belonging to the American Indian, based upon a treaty between the United States government and Indigenous people, in the the eighteen century, facing a strong resistance from the Indigenous people, claiming that, first of all, the land is being used by the oil company illegally, and then, a possible oil leak would affect the Missouri River, which would be jeopardizing the lives of the area's inhabitants, which would include the American Indians, and others. The first part of the problem is ownership, which is legally oriented, but the second part has to do with science, because, it deals with possible contamination of the Missouri River, which is the basis of life, and human survival. But, unfortunately, the issue is seen entirely as a political issue. This is another example of where science should step in and play a major role, with subordination of politics and economics. Science could be used as a very powerful tool to find some kind of economically oriented accommodation, that would be acceptable to all parties, involved. But, American style is to put science in the back burners, in favor of the powerful, and the Mafia of International Capitals. **With formal logic, we could find out how many Presidents we have had since the American Revolution, because, it is a matter of simple** mathematical operation, but to unlock the secrets of the Universe, involving unlimited dimensions, we need a more scientifically advanced form of reasoning, and that in my opinion is dialectical materialist logic, as a beginning point of an unlimited journey, which would make life much easier for all of us. With love and respect for all mankind, Hossein Dallalbashi, 07/29/2016.

Chapter 1
FUNDAMENTALS OF
DIALECTICAL MATERIALISM

DIALECTICAL MATERIALISM BEGINS BY ENTERTAINING what it calls the fundamental question of philosophy, which is to determine the relation- ship of mind to matter, the primacy of one over the other. By that, dia- lectical materialism wishes to establish the philosophical proposition that the Universe is made of matter, and while it recognizes the existence of mind (thoughts, ideas, conceptions, and concepts, it, therefore, categorically concludes that matter has always existed prior to the development of any conscious mind and its thinking activities. This not only excludes the existence of human conscious mind prior to the existence of nature, or what it calls material existence, but it also implies the impossibility of the existence of any material entity, meaning any divine, conscious being prior to the material existence of the Universe. It further expresses the concepts of un-creatability and indestructibility of the material Universe by any material things. The material Universe has always existed, and will continue to exist in various forms.

Central to dialectical materialism is the concept of the material Universe, being in ceaseless motion, which is to say that nature from the smallest particles to the biggest Planets are in constant movement. This has been incontrovertibly established as a material fact by modern natural sciences atomic physics, chemistry, biology and genetics, and others. This establishes the notion of inseparability of matter, and motion. In this regard, Engel maintains : " of the bodies that don't move there is nothing to talk about". This motion of matter is not continuously repetitious on eternal basis; it implies novelty in the movement of matter, which in turn elevates the concept of evolution of the matter to the center piece position of the understanding of material Universe. AT this juncture, the Universe is not only made of matter, but matter in uninterrupted motion. And further still, matter is constantly evolving in parts as well as in its entirety.

Another important part of dialectical materialism is its understanding of what it calls Universal connection. This Universe is not just a collection of individual parts totally unrelated, each a dance and song of its own. It is atomically integrated, not necessarily of the same magnitude and effects.

This evolving material Universe is not just moving mechanically, it evolves according to certain laws discovered from nature itself.

Even though, at this time, man seems to be co- existing with nature, as if their existence had always been simultaneous, he is, in fact, the evolutionary by- product of recent phenomenon as compared, for example with the emergence of our Solar System, approximately, some 5 billion years ago. Therefore, man historically, and materially evolved the means, the sense organs, needed to investigate the rest of material existence, its laws, movement, evolution, and so forth. To understand nature means to discover its laws inherent in nature itself. Nature exists independently of man and so do its laws. Truths about the material Universe mean discovering these laws very accurately.

There are three laws that are said to be of universal character: The first is the law of unity and opposition of opposites; [second], is the law of qualitative transition; [third], is the law of the negation of negation. Simply described, the law of unity, and opposition of opposites means that in order for things to evolve, they must first interact. This interaction is based upon parts of the opposing tendencies involved. They can only exist, and develop by interaction (unity), while maintaining their opposing tendencies (opposition) to various degrees.

The law of qualitative transition exposes the processes of give and take among these entities of opposing tendencies and that addition and subtraction of individual parts, while having an accumulative effect, would result in a qualitative transition of that object to a higher level.

The law of negation means any evolved process would be further evolved infinitely. Any evolution or development would be further evolved and developed.

It is against this intellectual background that Lenin tries to establish his theory of knowledge and concept of truth.

it is important to know that Lenin did not actually sit down to write a chapter on the theory of knowledge. One can establish a system of thoughts which he used in the form of arguments against his philosophical opponents of his time, which when pieced together, would become his systematic theory of knowledge. Volume 14, among Lenin's collected works, "Materialism and Empirio Criticism, written in 1908, conducting philosophical arguments against Russian and European philosophers, and natural scientists such as Bazarov, and Ernst Mach, to name two, represents such a collection of philosophical statements. Throughout this paper, I have used statements and passages contained in that volume to, first, support ideas that are attributed to Lenin's theory of knowledge, and then subject them to criticism based on Engels's teachings and my own ideas of what dialectical materialism should be.

Chapter 2
LENINIST THEORY OF KNOWLEDGE

(FUNDAMENTAL TO THE THEORY of knowledge (epistemology) is his philosophical attitude towards the relationship between nature on the one hand and human beings investigating nature on the other; or simply, the subject object relationship.

A much stronger emphasis is placed upon the proposition that nature is uncreated, has always existed, and will continue to exist not necessarily in the same forms, and independently of human beings' trying to understand it. There is nothing beyond this material universe, and that there is not anything that could not be known in time. If we did not admit the existence of material entities or the knowability of the material universe, then we would be lapsing into all forms of idealism.

Lenin states: "...The physical world exists independently of the mind of man and existed long prior to man, prior to any 'human experience'; the physical, the mind is the highest product of matter (i.e. the physical), it is the function of that particularly complex fragment of matter called the human brain." Lenin, Vol.14, pp. 227, & 228...

Emphasizing the primacy of nature over man, Lenin holds: "...In order to present the question in the only correct way, that is, from the dialectical materialist standpoint, we must ask: 'Do electrons, ether and objective realities outside the human mind or not?' The scientists will have to answer this question unhesitatingly; and they do invariably answer it in the affirmative, just as they unhesitatingly recognize that nature existed prior to man and prior to organic matter. Thus, the question is decided in favor of materialism, for the concept of matter, as we already stated, epistemologically implies nothing but objective reality existing independently of the human mind and reflected by it." Lenin, Vol. 14, p. 261.

Once Lenin establishes the relationship of man to nature, and correspondingly admits the primacy of the latter over the former, he then proceeds to explain the processes of acquiring knowledge of nature by man. **This endeavor for Lenin would remain unresolved, unless he categorically pinpoints the source of knowledge**. Because the way one responds to this question shows epistemologically whether he Is speaking from the idealist or materialist points of view.

Lenin claims: "All knowledge comes from experience, from sensation, from perception. That is true --- **but the question arises, does objective reality 'belong to perception', i.e. is it the source of perception**? If you answer yes, you are a materialist. If you answer no, you are inconsistent and will invariably arrive at subjectivism, agnosticism, irrespective of whether you deny the knowability of the thing in itself..." Lenin, Vol. 14, p.128. *It is my position that there is nothing that is knowable in its self. Everything is knowable through an integrated atomic inter- relationship*

Arguing against Machism, a Russian atomic physicist, Lenin holds: "But in fact, the Machists are subjectivists, and agnostics, for they do not sufficiently trust the evidence of our sense -organs, and are inconsistent in their sensationalism. **They do not recognize objective reality, independent of man, as the source of sensation, a true copy of this objective reality,** thereby coming into direct conflict with natural science and throwing the door open for fideism... **For materialists, our sensations are images of the sole and ultimate objective reality, ultimate not in the sense that it has already cognized to the end; but in the sense that there is not and cannot be any other...**" Lenin, Vol. 14, p.129.

After Lenin makes a definitive distinction between the nature (objective reality) as the source of knowledge and our sense-organs as the means of acquiring knowledge from nature, he wants to know how accurately/ faithfully the sense-organs, represented by our mind reflect nature and therefore correspond to it. **The reason for this is that the material universe, and its movement, and so forth, contains and therein represents the truth about itself, un-influenced in any form or fashion by our sense-organs to discover these truths**. Therefore, the question of the exact ref lection of nature as the most genuine truths and the reliability of human knowledge become very essential in Lenin's theory of knowledge.

It is said that truth is objective because everything about nature is self contained; and in time nature itself reveals itself by giving out its secrets to the human mind. Thus, objective nature, uninfluenced, unaffected by the human mind, is faithfully reflected. If our mind reflects nature exactly, then our knowledge of nature corresponds to the objective reality of nature.

On self containment, independence, and the objectivity of truth Lenin maintains: "If color Is a sensation only depending upon the retina (as natural science compels us to admit), then light rays, falling upon the retina, produce the sensation of color. This means that outside us, independently of our mind, there exists a movement of matter, let us say ether wave of a definite length and of a definite velocity, which, acting upon the retina, produces in man the sensation of a particular color. This is precisely how natural science regards it. It explains the sensation of various colors by the various lengths of light waves existing outside the human retina, outside man and independently of him. this is materialism… Lenin, Vol. 14, p.SS.

Arguing against Bogdonav, Lenin believes: "Natural science leaves no room for doubt that its assertion that earth existed prior to man is a truth. This is entirely compatible with the materialist theory of knowledge. The existence of the thing reflected independent of the ref lector (independence of external world from the mind) is the fundamental tenet of materialism. The assertion made by science that earth existed prior to man is an objective truth." Lenin, Vol. 14, pp.123, 124.

in further support of the proposition that man's views are reflections of the external world, Lenin considers: "...It is nevertheless beyond question that mechanics was a copy of real motions of moderate velocity, while the new physics is a copy of real motions of enormous velocity. *The recognition of theory") as a copy, as an approximate copy of objective reality, is materialism...* Lenin Vol. 14, p. 265. And finally, shedding more light upon the relationship of man to nature, and the theory of correspondence, or the theory of reflection Lenin concludes: "*...Matter is a philosophical category denoting the objective reality which is given to man by his sensations, which is copied, photographed, and reflected by our sensations, while existing Independently of them...*" Lenin, Vol. 14, p. 130.

Lenin's concept of truth, and the relationship of relative to absolute truth is based upon his theory of reflection, and correspondence. Basing his philosophy on the commitment towards "objectivity of truth", *Lenin tries to resolve another cognitive problem how does man cognize objective truth, at once, completely unconditionally, absolutely, or approximately relatively?*

The degree of closeness, and approximation of our ideas to objective truth (nature), the way "nature really is", unaffected by the inquirer demonstrates the relationship of relative to absolute truths. If the correspondence of our thoughts' to nature (objective truth) Is incomplete, partial, then we only have relative truth. *But, if the reflection Is exactly and accurately and completely copying nature, subsequently we have attained the Leninist "absolute truth."* So, from the position of relative truth to the attainment of absolute truth is a matter of degree and graduation.

Lenin goes on: "Two questions are obviously confused here: 1) Is there such a thing as objective truth, that Is, can human ideas have a content that does not depend on a subject, that does not depend either on a human being or on humanity

2) If so, can human ideas, which give expression to objective truth, express it a at one time, as a whole, unconditionally, absolutely, or approximately relatively *This second question is a question of the relation of absolute truth to relative truth.*" Lenin, Vol. 14, P. 133

Still on the attainability of absolute truth Lenin continues: "Human thought, then, by its very nature is capable of giving, and does give, absolute truth which is compounded of a sum total of relative truths. Each step in the development of science adds new grains to the sum of absolute truth, but the limits of the truth of each scientific proposition are relative, now expanding, now shrinking with the growth of knowledge..." Lenin, Vol. 14, p. 135. "The mastery of nature manifested in human practice is a result of an objective, absolute eternal truth." Lenin, Vol. 14, p. 190. *"That absolute truth results from the sum to- total of relative truths in the course of their development; that relative truths represent relatively faithful reflections of an object independent of mankind that these reflections become more and more faithful; not- withstanding its relative nature, [relative truth] contains an element of absolute truth* Lenin, Vol. 14, p. 309.

Lenin quotes Karl Kautsky: "...That I see green, red, and white is grounded in my faculty of sight... But that green is something different from red testifies to something that lies outside of me, to real differences between things.... The relations and differences between the things themselves revealed to me by the individual space and time concepts...**are real relations and differences of the external world, not conditioned by the nature of my perceptive faculty...•** Lenin, Vol. 14, p. 204.

Chapter 3
ENGELS'S THEORY OF KNOWLEDGE

THE SAME PROBLEM ATTRIBUTABLE TO Lenin is also associated With Engels's philosophical writings, in the sense that he too never specifically wrote on problems of theory of knowledge (epistemology). One can only, by detailed studies of his writings, extract materials contained in his philosophical arguments against his contemporary philosophers, social scientists **which when reconstructed, would form Engels theory of knowledge.**

This is what I have attempted to do in this paper, which Is to single out passages, statements, and quotations, perhaps most representative of his philosophical attitude, which I think, would form a kind of theory of knowledge contrary to the way his theory of knowledge is usually portrayed, and interpreted by some Marxist Leninist" followers.

To understand Engels's theory of knowledge, I feel one would have to start from the position of an evolutionary material universe as the very basis of Engels' epistemological analysis; more specifically, the way he considers motion, uninterrupted, inherent to, and inseparable from the material universe as the very basis of his theory of knowledge. To him, motion is the very basis of all forms of existence. For example, **it is stated that the entire universe (matter) is in constant, uninterrupted, ceaseless motion. By "matter," he means any object ranging from** the smallest particles up to the biggest planets. This raises the question of the "ultimate constituents" and the make- up of our external world. Engels states: "Physics, like astronomy before it, had arrived at a result that necessarily pointed to the eternal cycle of matter in motion as the ultimate reality." Engels, Dialectics of Nature, p. 11. He continues: "Motion in the most general sense, conceived as the mode of existence, the inherent attribute of matter, comprehends all changes of place right to thinking.... *Hence, in the historical evolution of natural sciences we see how, first of all, the theory of the simplest change of place, the mechanics of heavenly bodies, and terrestrial masses was developed; it was followed by the theory of molecular motion, physics, and immediately afterwards, almost along side of it and in some places in advance of it, the science of the motions of atom, chemistry.*", Dialectics of Nature, p. 35.

From the above statement, one could see that various sciences such as physics, chemistry, and biology, and other sciences derived from them, and their distinctions are based upon the study of different kinds of motions inherent in the objects of investigations. But these forms of motions, of molecules, atoms or even sub atomic structures, are not only inter- related and mutually determined, but are also inner related. They are inter related and mutually determinative because one form of motion, say, atomic is transferred and transformed into those of molecular, and macro body, and vice versa; but they are also inner determined, and mutually determinative because even the smallest parts of nature, i.e. atomic and sub atomic structures are not made of one aspect of the motion, i.e. negative or positive, south of north poles.

"The interaction of various parts of nature upon one another consti- tutes motion. The whole of nature accessible to us forms a system, an inter- connected totality of bodies, and by bodies we understand here all material existence extending from stars to atoms, indeed right to ether particles, in so far as one grants the existence of the last named. **In the fact that these bodies are interconnected is already included that they react on one another, and it is precisely this mutual reaction that constitutes motion." Engels, Dialectics of Nature, p. 36.**

No matter how far we go in dividing matter--- to the atomic or even sub atomic structures, we would still find a situation where the process of motion would still possess its inseparable opposite parts of motion, i.e. positive and negative. It is this very complicated process of motion which forms a "contradiction." by "contradiction," he simply means processes of motion. Engels goes on: "All motion consists in the interplay of attractions and repulsions.... **Hence, all attractions and repulsions in the universe mutually balance one another**....Dialectics has proved from the results of our experience of nature so far that all polar opposites in general are determined by the mutual actions of the two opposite poles on one another, that the separation and opposition of these two poles exists only within their unity and inter connection, and conversely, that their inter connection exists only in their separation, and their unity only In their opposition....How does motion present itself in the interaction of attraction and repulsion? We can best investigate this in the separate forms of motion itself." Engels, Dialectics of Nature, pp. 38 39

"...All innumerable operative causes in nature, which until then had a mysterious, inexplicable existence as called 'forces' mechanical force, heat, radiation (light and radiant heat), electricity, magnetism, the force of chemical combination and dissociation are now proved to be special forms, modes of existence of one and the same energy, i.e. motion." Ludwig Feuerbach And the Outcome of Classical German Philosophy, p. 66.

On the mutual penetration and determination of the constituent of motion, he further explains: **"...Two poles whose activities did not com- pletely compensate each other would not indeed be poles, and also have so far not been found in nature..." Engels, Dialectics of Nature, p. 44.**

Engels is concerned here about interaction of various parts of motions, not their positions. **"...In as much as attraction, repulsion compensate each other in the universe, and accordingly it would appear, a matter of indifference which side of relation is taken as positive and which as negative, just as it is of no importance in itself whether the** positive obscissae are counted to the right or the left of a point in a given line..." **Engels, Dialectics of Nature, p. 47.**

Rebuffing the notion that only one part of the process of motion plays an important, effective role while other parts are passive and receptive, he maintains*: "All natural processes are two sided, they rest on the reflection of at least two effective parts, action and reaction. The notion of force, however,* owing to its origin from the action of the human organism on the external world, further because **of terrestrial mechanics,   implies that one part is active, effective, the other part being passive, recep-tive... The reaction of the second part, on which the force works, appears at most as a passive reaction,** as **resistance....**This mode of conception is permissible in a number of fields even outside pure mechan-ics, namely where it Is a matter of the simple transference of motion and its quantitative calculation. But already in the more complicated physical processes it is no longer adequate, as Helmholtz's own examples prove." Engels, Dialectics of Nature, pp. 51 52

As a matter of fact, different types of matter are recognized by various kinds of motions and their complexity. Motions determine the nature of matter. He holds: "The different forms and varieties of matter themselves only to be recognized through motion, only in this are the properties of bodies exhibited; hence, the constitution of moving bodies results from the forms of motion...d" **Engels, Dialectics of Nature, p. 1 56. Engels, by agreeing with Hegel, arrived at the notion of a relative, material universe. "Motion of a single body does not exist, only relative mo-tion...." ibid, p. 1**

56. "The true nature of the determination of essence is expressed by Hegel **himself: "In essence everything is relative (i.e. positive and negative, which have meaning only in their relation, not each for Itself)." ibid, 161.**

This relativity of the material universe is based upon the relativity and mutual determination of motion. Engels continues: **"The first thing that strikes us in considering matter in motion is the interconnection of** the individual motions of separate bodies, their being determined by one another..." ibid, pp. 170,171.

Different kinds of motion, not only overlap one another, but they also transform into each other. **"Reciprocal actions is the first thing that we encounter when we consider matter in motion as a whole from the standpoint of modern natural science. We see a series of forms of motion, mechanical motion, heat, light, electricity, magnetism, chemical union, and decomposition, transitions of states of aggregation, organic life, all of which, if at present we still make an exception of organic life, pass into one another, mutually determine one another, one in one place cause and in another effect..."** ibid, p. 173.

From the position of the relativity of matter and motions, Engels arrives at mutual causation. He reports: "...Grove's whole misunderstanding about causality rests on the fact that he does not succeed in arriving at the category of reciprocal action.... only from this universal reciprocal action do we arrive at the real causal relation..." ibid, p. 174.

On the nature of motion Engels quotes Hegel: *"It is better expressed (as Thales does) that magnet has a soul, than that it has an attracting force; force is a property that, separable from matter, is imagined as a predicate. Soul, on the other hand, being its movement, is identical with the nature of matter...•* ibid, p. 184. *Engels concludes: "Life is therefore also a contradiction which constantly originates and resolves itself ...The unity of all motions in nature is no longer a philosophical assertion but a fact of natural science." Engels, German philosophy, p. 66. Here obviously Engels use the term "contradiction" as a process of motions.*

I have tried to show that despite this unfounded claim, at least in terms of epistemology, there are vast [diverging] differences between the two.

Chapter 4
A CRITIQUE OF LENIN'S THEORY OF KNOWLEDGE

IN LENIN'S THEORY OF KNOWLEDGE, all aspects of the processes of acquisition of knowledge are well defined, each having its unmistakable function to perform. For instance, nature (objective reality) self containing absolute truth, is the object of human investigation, and knowledge. Human being, possessing sense· organs to acquire truths from nature, is the subject. The object (nature) acts upon the subject (human sense organs} resulting in sensation, which in turn, uninfluenced by the subject is the ref lection of nature.   The closer the ref lection of nature, the nearer we get to "absolute truth," i.e. reflecting nature in its entirety. Partial reflection of nature is relative truth; whereas complete ref lection means "absolute truth."

What I will be doing in the following is to criticize Lenin's theory of knowledge, more specifically his theory of reflection and his concept of absolute truth. This critique Is firstly based upon Engels' understanding of motion as the basis of material existence, and secondly within the context of a book I recently wrote entitled, "Self Motioned Atomic Dialectical Materialist Logic, which 1 think not only supports Engels teaching of dialectical materialism, but extends it to Its logical conclusion.

It is usually claimed that Leninism is not only in line with Engels philosophical thoughts, but also represents their further evolution. In this paper,

Lenin himself, in his philosophical writings, makes references to Engels as an authority worthy of respect and theoretical emulation. But all his references are usually made to two of Engels' books, namely, "German Philosophy," and Anti Duhring." The book, "Dialectics of Nature," is noticeably absent. It is said that Dialectics of Nature was first published in 1926 and had therefore been unavailable in Lenin's time and certainly inaccessible to Lenin who died in 1924.

Dialectics of Nature is a collection of personal notes, statements, and quotations jotted down by Engels during almost a decade of detailed studies of natural sciences and mathematics after his retirement from business. The orientation of these personal notes, which were to have been used for a book on dialectical materialism, had Engels lived longer, is unmistakably that of natural sciences as opposed to his previous writings, used by Lenin, which could both be interpreted from political and natural science standpoints. Engels attitude towards natural sciences clarifies this point quite well: ...Secondly, Fuerbach **Is correct in asserting that, exclusively, natural scientific materialism was, indeed, the foundation of the edifice of human society knowledge, but not the building itself." Engels, German Philosophy, pp. 28 7 29. He continues: "...When I retired from business and transferred my home to London, thus enabling myself to give the necessary time to it, I went through as complete as possible a 'moulting,' as Liebig calls it, in mathematics and natural sciences, and spent the best part of eight years on it..." Engels, Anti Duhring, pp. 1 5 & 1 6.**

It is Engels as a natural scientist, and his corresponding theory of knowledge vs. Lenin's epistemology with its strong political orientation that I have tried to analyze. **In Lenin's theory of knowledge, the subject and object are taken as two separate entities with the latter existing independently of the former, while the subject is assigned the role of camera taking pictures of nature.** Nature (objective reality) is considered **very dynamic, changing, evolving, while passivity and recipiency, mirroring and copying are the characteristics of the subject. Lenin states: "Matter is a philosophical category denoting** the objective reality which is given to man by his sensation, which is copied and photographed and reflected by our sensations, while existing independently of them..." Lenin, Vol.14, p.130.

This, as Engels would say, is suggestive of mechanical materialism, because it ignores the dynamic interaction of [the] subject- object relationship, as two evolving overlapping parts of material existence while the emphasis is placed upon nature as the sole creator of sensation, and therefore, provider of knowledge. Lenin considers the effect, the product, the outcome of knowledge [to be] the cognitive process, instead of the causes of sensation and its corresponding material contributors.

Engels sees the problem quite differently, as an interactive approach: "If we change heat into mechanical motion or vice versa. It is not the quality altered while the quantity remains the same? Quite correct. But it **is with change of form of motion as with Heln's vice; anyone can be virtuous by himself, for vices two are always necessary. Change of form of motion is always a process that takes place between at least two bodies, of which one loses a definite quantity of motion of one quality.".** Engels, Dialectics of Nature, p. 28. **It is quite obvious that if man is a part of nature, having evolved from it, as Engels asserts and Lenin reminds us, then man and all his senses are materially composed, and as Engels would say, their material constitution is mutually determined by their motions. Therefore, the senses such as sight, the physiological apparatus of tasting, hearing and smelling, as well as the brain with its alleged billions of nerve cells, must be mutually determined.**

the modes of existence of these sense organs are based upon their motions. But being parts of nature, the determination of their motions go beyond their own material boundaries, and must necessarily interact with the rest of nature. **Engels testifies: "Further, we find upon closer investigation that the two poles of an antithesis, positive and negative, e.g. are as inseparable as they are opposed, and despite all their opposition, they are mutually interpenetrating. And we find, in like manner, that cause and effect are conceptions which only hold good** in their applications to individual cases; but as soon as we consider the Individual cases, in their general connection with the universe as a whole, they run into each other, and they become confounded when we contemplate that universal action and reaction in which causes and effects are eternally changing places, so that what is effect here and now will be cause there and then, and vice versa." Engels, Anti Duhring, p. 32.

On the interconnection of [the] human brain and its product (thought) with nature, Engels states: "...But further question Is raised what ·thought and consciousness really are and where they come from, it **becomes apparent that they are products of [the] human brain and that man himself is a product of nature, which has developed in and along with its environment; hence, it is self evident that the products of the human brain, being in the last analysis also products of nature, do not contradict the rest of nature's interconnections, but are in correspondence with them...**" Engels, Anti Duhring, p. 49.

Both the subject as well as the object must not be considered as finished products, as Lenin's theory of knowledge asserts, each living independently of one another, in exactly the same form. While it is true that nature existed prior to man as we know him, it is not true that nature had always existed in the same form. The emergence of man has changed [the] physio -chemical and biological environment to a great extent. But this is not the question; the question is can nature exist without and independently of him. **We have got to be insane to say no. Very few people would challenge the fact that man is of a later phenomenon.**

The real question ignored by Lenin is now that we have both man and nature, how do they interact? Because if they are both material, then their modes of existence have to be determined by their mutual motion, as Engels asserts. This complicated and mutually determined motion must be investigated on self motioned sub atomic, molecular, genetic, cellular, organismal, and environmental levels. Here is what Engels has to say on the mutual determination of all things: "...Motion in the cosmic space, mechanical motion of smaller masses on various celestial bodies, the vibration of molecules as heat or as electrical of magnetic currents, chemical disintegration or combination, organic life, - **at each moment individual atom of matter in the world is in one or other of these forms of motion, or in several forms at once... "Engels, Anti Duhring, p. 75.**

If this is the kind of universe we have where "...at each moment individual atom[s] of matter in the world [are] in one or [the] other,·.at once...", as Engels argues, knowing that all the sense organs of human beings are consisted of innumerable combinations of sub atomic, atomic, molecular, genetic structures, then how are these micro entities interacting with nature, what is the extent of their mutual determination of their existence, and to what extent are they dynamically participating in the birth or creation of images of the external world in our brain? **Lenin's response is very simple. Nature may be dynamic, evolving, but it alone acts on the subject, creating sensation.**

, sensation is a material process associated with nerves, and "protein substance. This sensation Is only a reflection of what is already there in nature, i.e. the "absolute truth." No interaction of [the] subject object relationship is permissible, recognizing that images are based upon this interaction as it could be deduced from Engels' understanding of motion as a mode of existence. Sensation according to Lenin is the effect, the result, the copy of nature. But to Engels" only to be different in various types of biological organisms.

Engels explains: "Not only all primitive animals, but also the plant animals, or at any rate the great majority of them, show no trace of a nerve apparatus. It is only from worms on that such apparatus is regularly found, and Herr Duhring is the first person to make assertion that those animals have no sensation because they have no nerves. Sensation is not necessarily associated with nerves, but undoubtedly with certain albuminous bodies (protein substances) which up to now have not been more precisely determined..." Engels, Anti Duhring, p. 98.

Further on the subject, Engels reinforces: "...If therefore, tree frogs and leaf eating insects are green, desert animals sandy yellow, and animals of the polar regions mainly snow white in color, **they have not certainly adopted these colors on purpose or in conformity with any ideas; on the contrary, the colors can only be explained on the basis of physical forces and chemical agents..."** Engels, Anti Duhring, p. 98.

"...Light and darkness are certainly the most conspicuous and definite opposites in nature; they have always served a rhetorical phrase from the time of the fourth gospel to the lumieres of religion and philosophy in the eighteenth century." Fick, p. 9. "The law long ago rigidly demonstrates in physics...that the form of motion called radiant heat is identical in all essential respects with the form of motion that we call light." Clerk Maxwell, p. 14. "These rays of radiant heat have all the physical properties of rays of light, and are capable of ref lection, etc. Some of the heat rays are identical with the rays of light, while other kinds of heat rays make no Impression upon our eyes. Hence, there exist dark light rays, and the famous opposition between light and darkness disappears from natural science in its absolute form. Incidentally, the deepest darkness, and the lightest· most glaring light have the same effect of dazzling our eyes, and so for us they are identical... The fact is, the sun's rays have different effects according to the length of the vibration, those with greatest wave length communicate heat, those with the shortest medium wave length, light, and those with the shortest wave length, chemical action (Secchi, p. 632 et Seq.), the maxima of the three actions being closely approximated, the inner minima of the outer group of rays, as regards their action coinciding within the light ray group. What is light and what is not light depends on the structure of the eye. Night animals may be able to see even a part, not of the heat rays, but of the chemical rays, since their eyes are adapted for shorter lengths than ours..." *Engels·, Dialectics of Nature, pp. 21 0 & 211. Engels sees light as lengths, vibration, and chemical action in- cluding the structure of the eyes, both to determine a given light, color, and so forth. Whereas to Lenin, even though color., are determined by the frequency of their vibration, the eye only "ref lects" what is already there, unaffected by the sense organs.*

Here Engels is describing the sensation as some kind of "protein substance" to be determined in the process of subject object interaction. Whereas to Lenin it is the effect, the result of [the] ref lection of nature. **Engels says to study effect instead of cause in a process would lead to mechanical materialism.**

In mechanics, the causes of motion are taken as given, and their origin is discarded, only their " effects being taken into account. Hence, if a cause of motion is termed a force, this does no damage to mechanics as such; but it becomes the custom to transfer this term also to physics, chemistry, and biology, and then confusion is inevitable." Engels, Dialectics of Nature, p. 55.

Lenin's theory of correspondence or reflection is based upon mechanical materialism. **Because if we accept, as he does, the attainability of absolute truth, the implication would be exhaustibility of [the] evolution of nature, at least as far as its constituents parts are concerned.**

We would reach a point where there would be nothing else to know because we would have attained absolute truth. Even though Lenin recognizes evolution, it is one with predetermined form where the evolutionary direction of the universe is quite clear. The result is a tacit acceptance of predestination of the universe with a teleological overtone. **Even the evolution of the new direction is already mapped out.**

Whereas in Engels understanding, the process of motion and therefore modes of existence are never the same. He goes on: **"...Motion itself is a contradiction. Even [a] simple mechanical change of position can only come about through a body being [at] one and the same moment of time in one place and in another place, being in one and the same place and also not in it. And the continuous origination and simultaneous solution of this contradiction is precisely what motion is...• Engels, Anti Duhring, p. 33.**

To Lenin, subject and object are two separate entities with the former mirroring, copying, or photographing the latter. It seems to me that our interaction with nature, even though taking place on self motioned sub atomic. molecular, genetic, cellular basis, as dismissed by Lenin, is also taking place on [the] organismal, or macro level. To Lenin, it does not matter on what level the interaction takes place, because we only take pictures of nature. **My own thinking' is that our interactions with nature, on the macro level, are much stronger at this stage of human biological evolution. Therefore; the mental images created and recorded in the brain are of macro magnitude and macro orientation. To interact with and have the brain record nature on the micro level, the human brain and other organs would have to await further evolution, acquiring the necessary natural equipment. the emergence and gradual development of the sixth sense is a testimony to this claim.**

Engels states: "...Each mental image of the world system is in actual fact limited, objectively by historical conditions, and subjectively by [the] physical and mental constitution of its originator." Anti Duhring, p. 50. He continues: "Dialectics, on the other hand, comprehends things and their representations, in their essential connection, concatenation, motion, origin, and ending." ibid, p.33. "An exact representation of the Universe, of its evolution, of the development of mankind, and of the ref lection of this evolution in the minds of men, can therefore only be obtained by the methods of dialectics with its constant regard to the innumerable actions and reactions of life and death, of progressive and retrogressive changes." ibid, p. 33. *"Dialectics, so called objective dialectics prevails through nature, and so called subjective dialectics, dialectical thought is only the reflex of the movement in opposites which asserts itself everywhere in nature, and which by continual conflict of the opposites and their final merging into one another, or into higher forms, determines the life of nature..."·* Engels, Dialectics of Nature, pp. 206 & 207.

My argument could very easily be misrepresented as the more traditional treatment of mind-matter relationship. This is not a situation in which mind gives order to nature, i.e. Kant. It is sense organs, including the brain, dynamically interacting with nature resulting In Images that are based upon this mutually determined material relationship.

Lenin takes what Engels calls images and representation of nature literally to be copies of nature. By doing so, he arrives at absolute truth. If the copies are inaccurate, partial, incomplete, "we have only relative truth". A sum total of these relative truths would eventually lead to "absolute truth."

If we take Engel's "motion as the mode of existence," and my evolving it into a more pronounced form, images as a result of the subject-object relationship, are we not implying that we are all different individuals having different interactions and therefore different ideas, knowledge and truths? This is exactly what we are saying. The bulk of humanity, at this point, is at a given biologically evolutionary level, with a minority embryonic ally developing the sixth sense. that is why, even though the interaction of every individual is different from other individuals, the variations are so nominal that they could very easily be dismissed as insignificant with regard to representing an alternative pattern of knowledge. There develops a pattern of "sameness" of mental images which may have basically the same relationship and meaning to all of **us in terms of our experiences of nature. The important thing to remember. is that every interaction dynamically, and atomically varies based upon** this relative interaction of the subject object relationship that we are being led to the relativity of existence and therefore, the relativity of truth.

"The **naturalistic conception of history, as found for instance to a greater or lesser extent in Draper and other scientists, as if nature exclusively reacts on man, and natural conditions everywhere exclusively determined his development, is therefore one-sided and forgets that man also reacts on nature, changing it and creating new** conditions of existence for himself." Engels, Dialectics of Nature, p. 1 72.

Engels **comments of motion as the basis of material existence: "Among natural scientists, motion is always as a matter of course taken as mechanical motion, change of place. This is a survival from the pre -chemical eighteenth century and makes a clear conception of the processes much more difficult. Motion, as applied to matter, is a change in general. From the same misunderstanding is derived also the craze to reduce everything to mechanical Grove is Strongly inclined to** believe that other affectations of matter.... are, and will be ultimately resolved into mode of motion,' p. i6 which obliterates the specific character of the other forms of motion. This is not to say that each of the higher forms of motion is not always necessarily connected with real mechanical (external or molecular) motion, just as the higher forms of motion simultaneously also produce other forms; chemical action is not possible without change of temperature, and electric changes, etc, but the presence of these subsidiary forms does not exhaust the essence of the main form in each case. **One day we will certainly "reduce thought experimentally to molecular and chemical motions in the brain but does that exhaust the essence of thought?"** ibid, p, 175

At one point, Lenin, in his book, "Materialism Emperio Criticism," gets very close to a subject object relationship position, instead of his usual ref lection theory of knowledge; but then backs away: "All boundaries in nature are conditional, movable and express the gradual approximation of our mind towards knowledge of matter." Vol. 141 p. 281. He continues: "Sensation is an image of matter in motion. **But through sensation, we only know nothing either of the forms of matter or of the forms motion; sensations are evoked by the action of matter in motion upon our organs. this is how science views it. the sensation of red ref lects ether vibrations of a frequency of approximately 450 trillion per second. the sensation of blue reflects ether vibrations of a frequency of approximately 620 trillions per second. The vibrations of the ether exist independently of our sensation."** Vol. 14, p. 30.

Notice that in above quotation, Lenin reveals that "but through sensations, we only know nothing either of the forms of matter, or forms of motions". Here he agrees with Engels, and assigns a role to sensation participating in the creation of knowledge. But, in the same sentence, he backs off, and switches back to his usual correspondence theory of knowledge. He goes on: "sensation is an image of matter in motion".

Lenin's theory of knowledge and the concept of absolute truth implies reducibility and fixity of the ultimate constituent parts of the Universe, **the ultimate sub-atomic particles remaining the same forever. If there are points beyond which one can't go, then knowledge about the object under investigation becomes exhaustible. This is how**

Lenin arrives at the concept of absolute truth. My own position is that the Unity and opposition of sub atomic structures always gives birth to newer constituent parts, and causes the disappearance of the old ones. Therefore, no ultimate constituent parts are ever attainable. Thus, no absolute truth could ever be achieved. Equating contradiction with the process of motion, Engels maintains that the process of motion is inexhaustible; and based upon this, he entertains the attainability of absolute truth: *Engels continues: "...With all philosophers it is precisely the 'system' which is perishable; and for the simple reason that it springs from a perishable desire of the human the desire to overcome all contradictions, once and for all, disposed of, then we shall have arrived at the called absolute truth: World history will be at an end..." Engels, German Philosophy, pp. 14, 15. d".. Engels continues: "Truth, the cognition of which is the business of philosophy, became in the hands of Hegel no longer an aggregate of finished dogmatic statements, which once discovered had merely to be learned by heart. Truth lay now in the process of cognition itself, in the long historical development of science, which mounts from lower to ever higher levels of knowledge without ever reaching by discovery called absolute truth..." ibid, p. 11. Engels continues: "... "The great basic thought that the world is not to be comprehended as a complex of ready made things, but as a complex of processes, in which the things apparently stable, no less than their mind images in our heads, the concepts, go through an uninterrupted change of coming into being and passing away.... But to acknowledge this fundamental thought in words and to apply it in reality in detail to each domain of investigation are two different things. If, however, investigation always proceeds from this standpoint, the demand for final and eternal truth ceases once and for all..." ibid, pp. 44 & 45.*

This shows beyond any shadow of any doubt that Engels's philosophical and scientific attitude is vastly different from Lenin's theory of correspondence, and theory of reflection in which we keep accumulating knowledge, piece by piece, with each piece representing

the "relative truth", and one day, the sum total of these pieces (relative truths) would arrive at "absolute truth".

Materiality and objectivity of nature on a macro body level is what Lenin's ref lection theory, the theory of correspondence relies upon, while it ignores the specificity of sub atomic, molecular, genetic structures of matter, and its relationship with atomic human beings on inter f lowing atomic levels. Whereas in ref lection theory, i.e. truth, in this case, a given color, is to be discovered partially, completely, and finally on absolute level; in contrast, to Engels' view, truth in a state of f lux is to be constantly born and reborn on a dynamic basis out of the actual conditions of individuals interacting with nature. In this, the individual as a complicated atomic biological organism participates in the dynamism of nature to determine the constant birth, and rebirth of the nature of the atomic truth. On this basis, Engels arrives at the concept of relativity of existence and, therefore, relativity of truth. This is not a vague academic, cultural, political, or intellectual argument. It is an existential issue that involves the inter atomic relationship of mankind with the rest of the Universe. Human beings are not above the entities, atomically detached, separate, and inde- pendent form nature, trying to investigate the Universe by taking pictures of it, and the more pictures they take, the closer they would get to knowing it, and one day, the pictures would pile up, each representing the "relative truth", as Lenin would consider them, and eventually, the sheer volume of the pictures would, one day, expose the sum total of absolute truths of the Universe. Even though taking pictures help us know things, **but to know things we have to atomically go through them, and every time we do, together with nature we disclose some aspects of the Universe**. We can't produce a baby human being, by taking pictures of the women we are with, or by writing love letters to them. We have to sleep with them, and that means interacting with them on millions of levels atomically, and genetically, in order, to make the conception and the birth of a baby possible. I could go to the library, for a decade, and read all kinds of books on how babies are made, and I would not make any progress towards making any babies, by any stretch of imagination, but it would take a few minutes to make them if we sexually, atomically interact with women. Human beings are just as much atomic, atomically integrated and interwoven with the Universe, and atomically evolving, as the rest of the Universe, and as conscious atomic beings atomically interacting with the Universe to disclose the secrets of the Universe. The term interaction could be confusing, if we interpret it as two different, separate, and independent entities, such as a human being interacting with nature to acquire knowledge. In Lenin's understanding, the human being uses his sense organs to interact with nature to establish certain truth. But, his sense organs do not do anything other than faithfully receiving information from nature, without being inf lluenced by nature, or influencing nature.

Chapter 5

MY ARGUMENT IS THAT HUMAN BEINGS, AND NATURE ARE ALREADY ATOMICALLY INTEGRATED, AND INTER-WOVEN BEFORE THEY INTERACT TO ACQUIRE KNOWLEDGE FROM NATURE

Lenin's teaching denies the atomic, and genetic integration of the two, before acquiring knowledge or truth from the Universe. The atomic interactions take place between our entire being, our brain, all our sense organs, all of our organs, as an organized unit with nature, while we are already atomically and genetically integrated with nature, and we may not be conscious of it. In fact, without this prior atomic integration, we could not even exist. Our bodies are be- inter-atomically influenced, modified by our atomic integration with nature, in every moment of our existence. Lenin wrote his books more than one hundred years ago, His understanding of dialectical materialism was based upon pre atomic physics, quantum mechanics, chemistry, organic chemistry, molecular biology, and genetics. He did not even have access to Dialectics of Nature, written by Engels, for which he spent 10 years of his life, when he retired from business, according to his own words, which was published in Moscow, in 1926, two years after Lenin had passed away. It is fair to say that Lenin's dialectical materialism is more profoundly, per- haps unconsciously, linked with Newtonian physics, or macro - body, pic- torialization of nature, or as Engels would say" mechanical materialism" understanding of the Universe, than modern advanced natural sciences. Never the less, Lenin should be given credit for his attempts to advance the importance of dialectical materialism, as a scientific logic, and provide leadership to the first historical social experimentation of attempting to create a socialist society, as crude, and distorted as it may have been. Some close friends of mine think that I have disrespect Lenin, by investigating the philosophical and scientific differences between Lenin, and Engels. To them I say: "forgive me if I have hurt your feelings, but I followed my conscience". I encourage others to do the same thing, to investigate, and criticize my work.

I wrote the book "self motioned atomic dialectical materialist logic" in 1979, and it was published by Vantage Press. Almost, in the same period, I wrote" a comparative analysis of Lenin's vs. Engels' theory of knowledge, (epistemology) a fascinating subject matter, without which, it would be impossible to understand any scientific approach. Years passed by, and I became involved in other issues in my life. While my book" self motioned atomic dialectical materialist logic" had been published, and was available publically, the comparative analysis of Lenin, and Engels theory of knowledge (epistemology) had been collecting dusts in my storage, and few people knew about, other than a few close friends. In the mean time, some unknown individual had reprinted my book, without my knowledge and consent, with no images on the cover, and has been selling it on Internet, for at least $50, a copy. The revival of the issues of "Democratic Socialism" by Senator Bernie Sanders's Presidential campaigns, and the positive response to Bernie's call, by the young people in U.S. encouraged me to republish my book, with CreateSpace Publishing company, and also add the comparative analysis of Lenin, and Engels theory of knowledge to it.

The ideas of socialism are not new to me, and from the early Sixties, up to the present, I have been involved with them on various levels, trying to find out what it is. One thing is very clear to me that the social experimentation in attempting to create a "socialist society" in Soviet Union, Eastern Europe, China, North Vietnam, Cambodia. North Korea, Cuba, and now Venezuela has been a failure, and a disappointment. These countries were not Marxist socialist. They were governmentally owned and operated economies, with some social services, provided to the general public. Soviet Union was disintegrated, China changed 180 degrees, and Vietnam, North Korea and Cuba will all be dismantled as soon as their traditional leaders pass away. Why did this happen? It happened for the simple, yet profound reason, that the working producing classes were not permitted to take over the factories, production facilities, and manage the economy on their own, in order to accomplish their historical role to transform the society, without any government interference. Instead governments confiscated the means of production, and a new class develop as professional bureaucrats, lifetime self assigned leaders, as new tyrants, and despots using the working class as wage slaves, as a single employer, and single despotic government, very much like capitalist countries. Marxist socialism defined governments as instruments of oppression of a dominant social class, against the rest of the toiling people. For that reason, Marxist socialism believed the working class in a socialist society must manage the production facilities without governments. Governments, and the called "workers political parties, must become ceremonial, and moral guidance in the beginning, and be completely removed from the society all together. For the simple reason that the issues are existential, and production management of the society, by the entire people involved in production facilities, and not politics, ideologies, religions, which are consisted of millions of forms of deceptions, fraud, manipulations, con artistry of the worst kind. In the Soviet Union, the government became a supper power, a single force to run the entire economy and the society. In China, the

29

government has become a global economic power, getting raw materials from the entire world, in exchange for selling them every product imaginable. This has completely destroyed the economies of the developing countries, not allowing them to develop their own industries, and also affecting the manufacturing sections of indus- trialized and hi tech countries. Many of the manufacturing companies in the U.S. Western Europe, Japan, Canada have moved to China, Vietnam, Bangladesh, Mexico, some other Latin American countries, in search of cheap raw materials, and very cheap labor. This process is irreversible, and not Bernie Sander, nor God himself can change it. It leads to third World War, if anybody attempts to change it. In North Korea, the same family has been militarily ruling, since world war two, very much like a dynasty. In Cuba, Fidel ruled until, he was no longer physically able to do it, then like a good king, he hand- picked his brother Raul, as the new President. In Venezuela, Hugo Chaves, hand -picked his successor, Nicolas Maduro, after he realized that he was dying.

It is my strong belief that no genuine Marxist socialist society could come about by any government. It is a mirage, a complete illusion. The capitalist economies globally, and the governmentally owned and run economies will survive for unknown future, because, the world population, very soon will reach 10 billion, and will out of hunger, desperation, and extreme poverty compete against one another in selling their labor power as cheaply as possible, to whomever, even the devils, in order to have any degree of employment, regardless of how little they would be paid for their labor. When a person is dying of hunger, with untold number of family poverty problems, and there are too many of them, and there are no possibilities to get employments, they would be happy and grateful for working for dirt cheap. Why? Because, the problem is existential, and survival is the issue, and not how much they would be paid by the hour, the day, the week, or the month. This is something that we do not understand in The West about people in developing countries, with very little, or no social entitlements, or absolutely vital social services. At this writing, the American Ford company is moving some of its production facilities to Mexico, to produce smaller cars. Some of our so- called progressives, including Donald Trump, from Right, are opposed to it, and calling it "a disgrace", because, it allegedly takes away jobs from the American working class. This is how the entire Chinese economy was industrialized, by capitals from the United States, Japan, Western Europe, and Canada. Without these outside massive Capitals, China would have remained a backward developing country. Even though, the labors of these countries are super exploited, but, it is undeniable that millions of jobless people are given employment, and brought to various industries.

To adopt a protectionist policy, recommended by Bernie Sanders, and even Donald Trump does not resolve the problem of unemployment and exploitation of labor. Global economy is one-sided, in favor of Capital. We have to struggle to globalize labor as well, meaning labors of all countries should go wherever there are jobs, on an international basis without any restrictions, as Capitals go beyond every boundary, without any problems. The capitalists and governmentally owned and run governments own the best real estates and trillions of dollars globally. They also have unlimited nuclear arsenals at their disposal to intimidate and destroy us, if they have to. So, their downfall very soon would be a day dreaming. There is no way that any force could overthrow them physically in the foreseeable future. Then, how does Bernie Sanders propose to create a "Democratic Socialism", against that background. Taxing the super rich more, and use that money to provide "free health care, and free college and university education, is not "Democratic Socialism". None of these things would happen in U.S. for the near future. Marxist Socialism, which has never been put into practice, would begin by introducing a plan of production, enabling millions of people to start producing for themselves, not for any governments, so literally hundreds of industries would begin to emerge and flourish, not by government, but by the entire labor force, managing factories, and collectively planning to expand all industries, providing employment for all able bodied individuals, men and women, indiscriminately. The ingenuity and wisdom, and productivity of people in charge of their own lives and productive activities would unleash the ability to produce material wealth, and abundance, beyond our belief. So, at that time, the working and producing classes would voluntarily set aside, and spend their own money on whatever personal and social issues they want to, including people contributed complete Universal health care, and going college and university education accessible to all on demand, to constantly retooling themselves educationally, in order to be prepared and ready for the new advancement in industries, and new scientific discoveries. These are not freebies from the government, or any gift from any benevolent successful influential, and rich social classes, it would be based upon people's sweat, and blood, and hard work. This is different from Bernie Sanders wanting to pass legislations, if the Congress would agree to, in forcing the multibillionaire, to pay more taxes to finance "Universal health care, and free college and university education. This may not be doable for the next twenty to thirty years. Then, what is the viable option for Bernie Sanders supporters now? What kind of future, would they have? The supper rich would take their capitals, and production facilities to other countries, where they would avoid paying taxes, all together, and employ dirt cheap labor, before they would allow Mr. Bernie Sanders spend their money for the general public. That is a day dreaming, to put it mildly. The Mafia of International Capitals are not waiting for Bernie Sander, and Donald Trump to make financial and economic decisions for them. In time, they remove Bernie Sanders and Trump from the scene, and replace them with more obedient politicians. When Bernie Sanders was campaigning for the Presidential nomination, I sent him a copy of my book" **Realize your dreams, produce for yourself** ", along with a letter of recommendation, to include to his rhetoric some of the ideas in my book, which I thought were useful for his "Democratic Socialism" audience. I mailed them to him. I have no idea whether or not he received them. But, I did not get any response from him. **The following is a copy of the letter I sent to Senator Bernie Sanders.**

Chapter 6

THE LETTER SENT TO SENATOR BERNIE SANDERS, U.S. PRESIDENTIAL CANDIDATE, ON 02/07/2016, A CRITIQUE OF HIS "DEMOCRATIC SOCIALISM"

Dear Mr. Bernie Sanders'

Thank you for retrieving the concept of socialism from the deepest swamp of taboos archives, and bringing it to the forefront, where some of its features could be exposed and discussed, at least among the young people, in this country. Leninism completely discredited Marxist socialism on a global basis, and mankind would suffer the consequences of that for an unknown future. Millions of people, around the world, have the idea that socialism means a complete government take over of all means of production, becoming a single employer of millions of new slaves. This was the horrible experience of Leninism, starting in the Soviet Union, being imposed upon the global Left for more than seventy five years. The remnants of this historical tragedy still remains in China, Vietnam, Cambodia, N. Korea, Cuba, and unfortunately in Venezuela, which is completely falling apart, while having billions of dollars of petroleum money, disgracefully wasted. They had to disintegrate with undesirable disgrace. They were destined to arrive at this point. Rosa Luxemburg, one of the sharpest Marxist theoreticians, in 1918, while in jail, no- ticed that Oct Russian Revolution was, in fact, Leninist, and not Marxist. Marxist socialism had envisioned a completely productive society, to be managed by productive people, from within, and not from above, by any form of govern- ments, of any kind, or by the homeless, the loafers, the lumpens, and all kinds of leaches. What you did was a phe- nomenal creative approach, which thousands, in the past, had failed to do. You have started from the consumption point of view of socialism, which is a redistribution, and conscious reallocation of collected taxes, to be spent on the most sensitive, and in the long run, productive priori- ties, such as tax based universal education, an universal health care for all, something that individuals, alone can not do, everybody chipping in to make it a reality. The Right claims that the general population is looking for a handout from the Government. The Government is a thief that steals from the public, and does not have any money of its own to give free lunches to the people. Collected taxes are people's money, and it is about time that people establish their own priorities, which are Universal edu- cation, and Universal health care. Building any army to carpet bomb the hell of people around the world is not the first priority. We have to have healthy and educated persons to function as human beings, and not sick and stupid ignorant people in our society. In today's standard, this is the minimum. The hell with Government. We could establish a people corporation and deposit our money in a given account and allow the corporation to buy educa- tion and health care, for us, instead of Government. If Government could take our money and as a single buy- er payer buy all kind of arms of mass destruction, U.S. style, "efficiently", then, the same Government could use our tax money to buy Universal education, and Universal health care, after they steal a good portion of it. This would be a tremendous boost to the economy, unleashing people's efforts to higher hope, higher production, and higher consumption, never to have been experienced be- fore. What is lacking in your program is the introduction of a feature that would entertain the idea of the working class, starting their own independent production facilities, independent from the Capitalist production, where in the long run, the working producing classes would gradu- ally deliver itself from the position of being wage slaves, working itself through becoming independent producers, while Capitalist production, and Governmentally owned and run economies would simultaneously survive, but on dimin- ishing forms. If you follow this financial and production policies, after the end of your second term, we would have 20 per cent of the economy owned and produced by the working class as partners. Then, you could boast of your "democratic socialism". Without this you probably would fade away. In fact, if you start talking about this right now, you go way beyond Hillary, right into the White House. By the way, this is the first time I vote, and I vote for Bernie Sanders.

I have written a book called "Realize your dreams, pro- duce for yourself". It is a blue print for the producing classes of establishing their own production facilities, in- dependent from the Capitalist and Governmentally owned and run economies, such as China, North Korea, Cuba and Vietnam. The book encourages the working class to establish group owned and managed production facilities, making their own products and services, and after having paid for the expenses, share the fruits of their own la- bor, all as partners, and not as employers, and employees. I am advising them to sell their labor powers to any entity, including all Governments, as a matter of last resort, in a do or die situation. Start telling people, that if your are elected, you establish two different Federal offices, di- rectly under your own supervision. The first one would start loaning money to people, who as a group, want to engage in some kind of production facilities, and services, and loan them directly from the Federal Government, with 2 percent for ten years. Then, establish another of- fice, providing active professional production and finan- cial consultation, free of charge, to make sure the loans are well spent. Don't leave them at the mercy of the banks, who receive unlimited phony printed money, from Federal Reserve, another rip off organization, at almost zero percentage, who turn around, and instead of mak- ing it available to the general public, they make it available to the so called "credit worthy", the multi millionaires, and billionaires, to buy all kinds of real estates, and production facilities, in U.S. and abroad. I am sending you a copy of my book for your studies, perhaps, you could use some of the ideas of this book in your speeches, and hopefully if you are elected, use them in your economic, and financial policies.

Mr. Sanders. I wish you good health, great stamina, and long life, to promote the divine causes of the unprotect- ed working classes, to become independent producers, so the Right wont claim that your are redistributing the Rich's money to "finance, free health care, and free edu- cation for the poor". This is the kind of economics they teach in the colleges, and universities of this country. I know that first hand, because, I sat through their classes for so many years. So, it is not our educational system that determine what the Wall Street is doing. It is the the Wall Street that tells our colleges and universities, what to write and what to teach. Thanks. JP

The end of my letter to Senator Bernie Sanders, sent to him on 02/07/ 2016, while he was still campaigning for the presidency of the United States. I did not receive any response, either from him, or his office.

Chapter 7
AN ALTERNATIVE PLAN OF ACTION FOR BERNIE SANDERS'S SUPPORTERS

I wrote a book called *"realize your dreams, produce for yourself "*, which is available at Amazon. Com. This book should have been alternatively called, "Alternative Production Economy". I highly recommend that the supporters of Bernie Sanders order the book and read it; and once they have done that, go to Amazon. com, lick on my name, on the customer review section, and give me some feedback as to what you think of the book. The content of this book is not an American soap opera novel, but are related to life determining issues, that have been discussed by a good portion of socially conscious, concerned individuals around the world.

In that book, I have dealt with the issue of the working producing classes, manufacturing and also high tech, not only in the United States, but also globally, to establish their own independent production facilities system, in a simple terms, establish their own businesses, in order to emancipate themselves from the historical wage slavery, forced hunger, poverty, the capitalist system induced and created unemployment, under employment and unspeakable poverty, imposed upon them, since I have no faith in capitalist and governmentally owned and run economies, such as the soviet Union, " the socialist community" in Eastern Europe,

China, North Korea, Cuba, Vietnam, Cambodia, and now Venezuela, to genuinely resolve the emancipation of wage slavery, this monumental life determining issues of global working classes, once and for all. The first thing we have to realize is that, the working class problems are not political, ideological, and religious issues in nature. They are existential, and production related the production of millions of products, and the scientific management of nature, and human being issues. The problem of why we produce, where we produce, for whom we produce, and who are the main recipients of our productions, are the main issues. **That means establishing political parties, trade unions, and other political institutions, and helping to improve the political system. will not help to solve the producing class issues, and all political approaches to supposedly bring about a new environment, simply are the social games of the oppressors, the tyrants, the imposters, the social leeches, the exploiters, the con artists, of thousands faces, and characters.**

Their con artistry is so multifaceted and refined, that the simple hearted hard working people are no match for them. These political and ideological strategies, and tactics have been practiced for the last several hundred years of modern capitalist existence, and they have worked in favor of the dominant social classes, and there is nothing in that for the working classes, other than reinforcing their wage slavery in a new forms, by cleverly crafting literally thousands of ideologically and politically side issues, that have nothing to do the working class destiny, are different forms of never ending entertainments, deepening our miserable lives. To use a political approach to make a transformation in the society has been the games of ruling owning classes. **This approach does not work for the working classes, because the working classes are deprived of land ownership, production facilities ownership, and therefore suffer from the absence of overall economic powers. Meaningful political power emanates from ownership of the means of production, land, labor, and capital, factories, banks, raw materials, and ownership of everything in general, houses, apartment houses, business properties, gold, silver, antiques, everything tangible, and not from**

possession of certain ideologies, political philosophies, and reading "good books". The working class approach must be existential, and production related, on global basis, and not politically and ideologically oriented, from a narrow-minded nationalistic level. This can only work on global levels, avoiding local, national arrow-mindedness, and a protectionist approach to trade. Therefore, the global working classes must completely abandon political approaches of all kinds, in order to completely liberate themselves from the position of international wage slavery, and concentrate on completely joint cooperative ownership of all means of production, that they use to produce, in order to make a an honorable, dignified, and independent living, and achieve all their dreams through group cooperative joint ownership, joint production management, and engage in global marketing of their products, and services, based upon private property principles, global market mechanism, and a completely global free trade, and competitive price system, and not rationing, acting as free as a bird. So far, the global working classes have been encouraged and motivated to only sell their labor powers to other social classes, in exchange for certain wages, and the only dream of making improvement in their lives would be, through trade unions and political activities, periodically asking their employers, for a raise. **This begging mentality of expecting other social classes to show mercy and compassion that they don't have, in order to give them employment, through all kinds of humiliation that must end, once and for all.** Owning properties, houses, apartments, places of business, buying properties for future security, and owning the means of production were attributed to dominant social classes, and it was something that the working classes should stay away from. It was undesirable for the working classes to own properties. So, while the working classes possessed the ability to produce abundance for others, they were encouraged to play the role of a beggar mentality, constantly begging the the businesses to give them employments, so they could keep producing for other social classes, in exchange for certain wages. They were promised by some of their Left leaders to wait, tolerating the inhumane, and unbearable situations, under capitalist economies, until one day, a working class government would emerge and liberate them from that status quo, and they would be working under a government, where all mean or production would be "socially owned", as opposed to the means of production, being privately owned under a capitalist economy, the working class, producing for itself. That was the greatest lie, ever told, for more than one hundred and fifty years. **That dream government was born and the means of production did not become "socially owned", but instead, there emerged a despotic government, a new economic power, in the hands of a group of self assigned professional life- time faceless bureaucrats, with the working classes, remaining as wage slaves**. Many individuals, in the genuine Left, do not agree with this analysis, and completely deny this altogether. That is fine. Let them remain as day dreaming, and ignorant until domes day. No socialism of Soviet model will ever take place on this Planet Earth, until the problem of socialization of the means of production, under complete production management of the entire producing classes, without any kind of commanding governments, is resolved, once and for all. The role of government, in the beginning, must be ceremonial, receiving the heads of other nations, and protective of the new production principles, making sure that the working class will be able to perform its historical mission of converting the society to one that is based upon production by all, and sharing the fruit of their labor by all. **This mission was not accomplished under Soviet type economy, because, the government became a replacement for the working class.** This has been the case with the Soviet Union, Eastern Europe, China, Vietnam, Cambodia, North Korea, Cuba, and now, Venezuela. In the sixties, we thought that if we were to have progressive minority mayors, police chiefs, governors, senators, or representatives, on the States levels, and on the Congress of the United States, we could create meaningful changes, that would result in fundamental transformation of the economic system, in favor of the American working classes. We succeeded in having all that, including having a Black President, Barack Hussein Obama, and nothing meaningful has come about in changing the fate of the working class. **That shows how naïve we were.** The minority politicians became as corrupt as the White race politicians. This is the nature of the beast. We form, and join trade unions, and have them negotiate a few more dollars, added to our wages, only to be offset, and overwhelmed by higher prices, in basic staples, in what we purchase, and consume, from the businesses we work for.

The working classes are born poor, and will die poorer. Those of us who are fortunate enough to have employment, and receive enough financial compensation for our work, so as to sustain our bodies, with minimum nutrition, so that the next day, we would be able to keep producing for other social classes. **They use us, abuse us, exploit us, dump us, fire us, and if we insist on better wages, they take their entire factories to other countries, where they don't pay taxes, hire the cheapest labor, causing forced unemployment for millions of workers, and we are not able to do anything about. As Albert Einstein said in defining insanity: "people keep doing the same thing, over and over again, expecting to have different outcomes".** We engage in meaningless political activities, in order to get a few dollars, added to our wages, and we never learn. The minority mayors, police chiefs, senators, and the House representatives, and even the Black President turned out to be as much of a crook as the White-oriented government officials. Yes, as Albert Einstein said: "we keep doing the same things, over and over again, expecting different results". **We can't in good conscience blame the ruling classes for our permanent dormancy, and failure to wake up, and take charge of our lives, and modify our destiny altogether, based upon or own interest, and not those of the other social classes.** If we keep doing the same thing, perhaps, we deserve it. The other day, I was listening to Ralph Nader show, on KPFK radio station. Mr. Nader and his quests were talking about the undesirable surprise of having Donald Trump, becoming the U.S. President, even though Hillary Clinton won the popular votes, by two million votes, while the oppressive, and deceitful electoral college system, the work of the American forefathers, creating a deceitful system of keeping the American political system under the control of the super rich, showing a middle finger to

American electorate, gave the presidency to Donald Trump. **If a electoral college system were to be practiced in Middle Eastern countries, which is common in most of them, in different forms, the United States would be mocking them as underdeveloped barbarians**. But, the U.S. practices that, with great pride, and they think it is quite normal, because, that oppressive system, an anti - ordinary people measure would prevent the ascendency of an "undesirable individuals" to the presidency of the United States. **Mr. Ralph Nader posed a question** to his quests on the show, that if somebody would donate 5 billion dollars to them, what would they do with that money in terms of improving the social conditions for the progressive causes, to prevent future takeover of the U S presidency by the Right Wing Republicans? **Each provided an answer, that was in line with improving the political conditions,** including hiring full social activists, to popularize the messages of progressive social causes, so that in the next presidential elections, the Democratic, or progressive candidates would be elected to various political offices, in order to improve the living conditions of the working class in the United States. **But, not even one of them suggested that a portion of that 5 billion dollars should be spent on exposing and finally eradicating from the political scene, and the US Constitution, the savagery, and the deceptive anti human beings concept of electoral college system, an old measure, built up in the US Constitution, by the landed aristocracy, the American forefathers, to safeguard their own socio - economic status, and monopoly, as landed slave owners, in order to prevent the ordinary people** from exercising their human rights, and elevating themselves to ruling their own affairs, putting their own desired people in office, the same system, that put Donald Trump into office. Here is what I would do if I had access to 5 billion dollars. I would not spend a dime of that money on any political issues. 5 billion dollars would be as divine as the Catholic Church, and it must not be consumed. **I would establish a bank, independent from the existing banking system, called "own manage produce& share, as partners Bank.**

I would loan up to 50 thousand dollars to a party of three to five individuals, without collaterals, who had a plan of establishing a small company, either producing some products, or providing some services that are needed in the society. I would hire a group of business experts specialists, in order to guide the small businesses, to unwavering success. This guiding group would constantly provide update information on the success or lack of it, in the operations. The Bank would not charge any interest for the first two years. After that period, there would be a monthly small return of the loan, plus three percent interest to the Bank. If the Bank loans some fifty thousand dollars to each group borrowers, on small projects, we would be able to create 100, 000 small businesses, manufacturing, hi tech, and service industries. The principals and the interests would be coming back to the Bank, with the 5 billion dollars being recycled, for creation of other small businesses, without being consumed. That is what I call "own manage produce &share, as partners economy. If people have their means of economic and financial survival, and they don't have to beg jobs from big businesses, and the governments, who cares what kind of con artists, whether Democrats, or Republicans, will be the next president of any country, including the United States. In twenty years time, a good twenty percent of the population would be engaged in "own manage produce, & share economy, which would definitely be a challenge to, both the existing capitalist economies of the Western countries, including the United States, and the governmentally owned and run economies, of the so called "socialist economies", China, North Korea, Vietnam, Cambodia, Cuba, and Venezuela. In the following pages, I am going to outline the highlights of my book, "realize your dreams, produce for yourself ", or "Alternative Production Economy", or "own manage produce &share, as partners economy" We have two major economic systems, that prevail on a global basis. They are: the capitalist system, such as the Western World, United States, entire Europe, Canada, Japan; and the governmentally owned, and run economies, such as the ex Soviet Union, Eastern Europe, China,

Vietnam, North Korea, Cambodia, Cuba, and now Venezuela. These countries claim to have established "Marxist Socialism", **but it is not true.** Why? Because, Marxist socialism means the entire working class managing all the means of production **without government.** Government was supposed to have been a ceremonial and moral pro- tective entity, first, and then in time be phased out completely. The means of production was supposed to have become **"socialized"**, a collective possession of the producers. But, instead, the means of productions **became governmentalized**, in the hands of the Communist Party, the Central Committee, and the self appointed non-elected bureaucrats, functioning as despots, most of them dying in office, in favor of other tyrants, **employing the worker class as wage slaves.** So, as you see the fate of the working classes did not change, they only exchanged one class of employers, in the capitalist economy, in favor of a single employer of the governmentally owned and run economy. **Marxism defined Governments as an instrument of oppression of one dominant social class against the general working population.** Would Governments in state owned and run economies be any different from those in the Capitalist countries? **Far from it! Governments are Governments in any type of economies**. They claim that their Governments are **"workers Government". Governments of any kinds, and Marxist Socialism don't go together.** Soviet Union had more than seventy years to first establish a ceremonial, morally protective Government, and then phase it out completely, leaving the entire management of the economy to the working class, **in order to complete its historical mission**. Up to its last day of its existence, it not only did not phase out, but instead became one the greatest military power controlling the economy, and everything else in the society, rightfully receiving the name of **a "totalitarian regime"**, with the **working class still remaining as wage slaves,** receiving some social services. No Government could establish any Marxist Socialism. It is only the historical job of the physical and mental high tech workers. How could government introduce "democratic socialism"? It is **a complete lie.** Socialism starts with production facilities managed by the productive working class. Democracy is a set of rules, used by different political parties, in a class divided society, to take turns, in expressing their narrow economic and financial interests. **If the working class produces for itself, how could it be more democratic to itself.** We use the word" democratic", if one social class is allowing another to express its ignored interests. The whole concept of democracy would fall apart, as soon as we start producing for ourselves, without employing others, and there is no third party, such as government imposing itself upon the production ownership and management of the society. If we all produce cooperatively, as partners, and managers of the means of production, then how could we be more democratic to our- selves. We would not be interested in redistribution of anything among ourselves. Redistribution of wealth in a Capitalist economy take place, because, some social classes unfairly receive more money than other social classes, then, we would distribute wealth to more disadvantaged social classes, in order to make it fair, and also prevent a violent social revolution. True Marxist socialism means working classes collectively managing their own production facilities, and fairness is implied in their collaborative coop- erative production activities, **and there is no need for democracy. So, for that reason "democratic socialism" is false concept in Marxist socialism.** But if, by socialism, we mean government ownership of the means of production, and arbitrarily employing workers in exchange for wages and some social services, then we could use the concept of "democratic socialism", as a tool for the working classes to have a greater involvement in the the governmentally owned and run economy, in that sense, the working class is struggling to have a greater involvement in the decision making processes. **So then, we are talking about two different types of socialisms, the Marxist socialism, where the workers collaboratively and cooperatively own and manage the production facilities, without governments, and the so- called governmentally owned and run economies, employing workers as wage slaves.** I think, that Senator Bernie Sanders is talking about the governmentally owned and run economies, impersonating itself as "socialism. **Forcing the super rich to pay more taxes, in order to finance Universal health care, and Universal education for**

the general population is not "democratic socialism". Because, in the West, and the United States the means of production are primarily owned by the capitalist classes, or the corporations, and we cannot use the concept of democratic socialism to a capitalist economy. This is more like the European social democracy than socialism. Every concept of socialism that does not begin with production facilities, being completely managed by the working class without government interference, is a **fraudulent claim,** whether or not we attach fancy terms, such as "democratic "to it. These claims are **empty election promises, that will never be realized,** fading into oblivion after the elections are over, and candidates occupy the offices. Even if Mr. Sanders had it approved by the Congress, which would have been very doubtful, the capitalists would have taken their production facilities to other countries, where they would never pay taxes, and the labor cost would be one tenth of what they pay in the United States. As long as there are plenty cheap lands, and more than seven billion persons to choose from, the capitalists do not feel obligated to any country or boundaries. **Their country is the entire Planet Earth,** in which to establish their production facilities. Their formula is "minimize costs, and maximize profits", which does not involve morality, ethical values, adherence to any religious teachings, feelings of patriotism, humanitarian commitments, values that they expect the general working population to strictly adhere to, and honor.

The global working classes have only one alternative available to them, and that is to establish their own economic system, within the womb of the existing capitalist, and government economies. I call that "alternative production economy", or **"research-own - manage produce- market, and share, as partners economy"**. How?

The global working classes should sell their labor powers to the capitalist, and governmental economies, and all governments, **as a last resort,** for the unemployment, atrocious, and intolerable poverty, and a roller coaster life they have imposed upon all of us. We would be managing our lives, with tranquility, peace of mind, without the fear of being laid off, in an un- timely manner, which would disrupt our lives, without the ability to repair. We would be our own boss. **America is ideal for this alternative production economy**, because, the labor power is very educated, and advanced, to enage in self managed businesses. **You would become crises proof, recession proof,** because, the economic control of our lives is in the palms of our own hands, not in the vicious irresponsible whimsical hands of a group of ruthless government bureaucrats and and capitalists, **who in order to punish us would take our jobs to China, and other countries. This is not a joke; it is a complete human liberation on all levels. Once we are economically independent, every other liberations would automatically follow. If we have our jobs, our businesses, our own houses, and we own everything, we could not care less what other people think of us. Once the economic basis of racism is resolved through the ownership of places of business, our houses, then other kinds of racism would become meaningless, and unimportant.** We should learn from our Jewish friends. From the time they are born, it is instilled upon their minds and hearts that they must begin to acquire everything, their houses, their places of business, gold and silver, and should never work for anybody else. I have a lot of Jewish friends in Downtown Los Angeles, **and I have never seen any Jew who would work for others**. They all have their own businesses. In the United States, from childhood, they keep telling us to go to school, become educated, so that we would get a job at a company. **We have to reverse this trend,** and teach our children to prepare a life of self employment and acquiring our own businesses, our own houses, our own apartments, and concentrate on ownership of every- thing in life, very much like what the Jews have been doing for the last five thousand years. **While they quietly keep acquiring everything in life, they encourage us to be good reliable workers. That is how stupid we are. We are born poor, and die poorer, and we never wake up, and recover from this profound stupidity and self imposed ignorance**. We have nobody to blame, but ourselves. That is a good place to start,

We must promote this global working class economy on the world wide level. We must establish information offices in every

state, country, and Continent to accumulate and disseminate business opportunities, resource availabilities on an international level. It is very important for this alternative production economy to become a global phenomenon, engaging in completely free trade, promoting international trade and brotherhood, and international intermarriage, that create an emotional and greater cultural bonds and understanding, among different nations.

This, in time, would create a greater close relationship among different nations, making the **boundaries obsolete,** and a genuine unification of all nations possible. Unification does not mean a **centralized entity ruling the Planet Earth.** It means removing all barriers, in order to have a harmonized life for all human beings. It means self management of life for all human beings without governments. **Each city, each neighborhood managing their own affairs, with free f low of humans, talents, wealth to all corners of the world.** I have no doubts that all of these things are achievable. We have to give ourselves a chance of **setting aside our fears,** and put it into practice. Obviously, **the global slave masters would say that these are day dreaming ideas,** and that there will be chaos, and that a military imposed separation is the best policy, as we have had it so far. This is an excellent approach to bypass the domination, and the slavery, that they have created for all of us. We can not fight them, because they have the ownership of the best production facilities, the oppressive forces of all governments, the army, the police, the educational system, the judiciary, and all forces of global domination and control, but we could by pass them by creating a separate Global working class economy without relying on any governments, and gradually become a formidable production and economic power, on an international levels. This is definitely achievable. Our economy would be a non-political non-ideological and non-religious entity based upon hard work, growing system of productions, of literally millions of products, **with humanity, decency, mutual respect, and science as a common de- nominators,** if we understand the evils of governments of all kinds, if we understand the the bondage, the oppression, and slavery of all kinds of governments, that have been self perpetuating in hundreds of forms for the last several thousand years of recorded history. Our global working class economy would create the conditions for our **complete human liberation**, in general. Once we liberate ourselves from wage slavery and establish our own forms of economic survival, every other form of injustice, oppression, dominance, and even every form of ugly racism would gradually diminish, paving the way for true human liberation. **There would be no need for emperors, kings, presidents, prime ministers, mayors, managers, supervisors, and bosses of all kinds**. There will be only human beings managing their lives, as they see fit. This is what I call the global working class economy, or what I call "alternative production economy", that I have explained in detail in my book" **Realize your dreams, produce for yourself** ". To get this idea off the ground, we would be encountering some difficulties, but nothing beyond the sufferings that we are used to, by being wage slaves, working for other dominant social classes, and abusive government, for centuries. But, in time the global working class would find harmonizing solutions for all problems. But, we have to be in charge of our lives, **there is no other genuine option.**

One problem we have at this time would be not having access to working capital. The Bernie Sanders supporters gave small donations of $ 27 each person, and it led to millions of dollars for the campaign expenses. The supporters of global working class economy, and I imagine Bernie Sanders supporters are included, could start thinking about establishing a **"global working class economy bank". Those of us who have banking experiences could volunteer, making and Ad hoc committee, and ask for donations from all interested individuals, to be deposited in a special designated bank account, and once we have sufficient funds, establish that bank.**

The supporters of global working class economy would start depositing their monies and savings in that bank, instead of depositing their monies in Bank of America, Wells Fargo, Chase Banks, that turn around and **give it to the Rich to become richer,** because, if we applied for a loan, we would not be "credit worthy". This is how they keep us in **forced**

poverty, having to accept whatever terms they impose upon us. They use our hard won monies to buy everything that constantly produce more capitals, for global domination. **And we are in deep coma**, functioning and behaving as good reliable faithful wage slaves. The global working class economy bank would start making small loans, as working capitals to group -initiated projects individuals, so that they could make the dreams of their life, owning and managing a business of productions, or services, dignity - creating reality. Some of us are loners, and would rather manage our own businesses alone, as a single person. **There is no problem**, you could do that. You could even use the members of your families, towards that end. That is fine; go ahead and do it. But, if a business project requires more than one person, then start looking for partners, and team works. **We should have a special magazine, collecting information on partnership opportunities, and share it with interested people**. Social media, face book, and You Tube, and others could make us closer together.

Marxist socialism is consisted of two stages: the socialist stage, in which each person would receive from what they produce, in accordance with their contribution to the economy; and the second stage, in which there is no money involved, and producers, regardless of their abilities, and contribution to the production facilities, would receive in accordance with their needs. Imagine, these should have taken place without government. Maybe, in the beginning, there would have a need for ceremonial and morally protective Governments, but, in the process, the governments had to have been completely phased out. This never took place, instead governments confiscated all means of production and used the working class as wage slaves. It became so unbearable that governments acquired the name **"totalitarian"**, because, they not only owned everything, but they controlled the entire society. **The historical mission of the working class to liberate itself and the rest of mankind became a farce**. That is why The ex Soviet Union, after seventy years of world domination, fell apart, **like a deck of cards**, China become a super capitalist power, producing **products for the entire world**. But, they did it with capitals, accumulated at the expense of the global working classes, in the United States, Europe, Japan, and Canada. Now, the manufacturing working classes, in the highly industrialized countries, are under- employed while China is producing millions of products, **with slave labor, paying less than two dollars,** an hour to their workers. They have the audacity of calling themselves Marxist socialism.

There is a great deal we could learn from true Marxist socialism. First of all we could **forget about the second stage** of Marxist socialism, which is the **Communist stage**, in which regardless of what our contribution is to the economy, we would be receiving, in accordance with **our needs. I honestly don't believe this is achievable,** so we would leave that for the future generation, to see if they could practice this. Human beings, as we know today, are not structured to be that **faire minded**. Maybe, in the future, we would consciously genetically modify human beings, so that they would act as decent human beings. **At this time, this does not exist, and we can not base our arguments on something that does not exist. I personally believe this conclusion is anti-dialectical, and must be eliminated.**

The best we could learn from Marx, and Engels, Marxist socialism is the prime indisputable importance of the processes of production and exchange, **without relying on governments**. History shows that governments **can not create socialism**. They only produce despotic regimes, giving us a few crumbs, in exchange for enslaving us completely. This is **no longer an option**. So, then, how do we liberate ourselves from all governments, wage slavery, and establish a life for ourselves?

The global rip- off systems of capitalist, and the governmentally owned and run economies **would last for a long time,** until history would make them irrelevant, and obsolete, but, this would not happen for a foreseeable future, because of their economic powers. **Then, what is way out for the Global working class?**

The answer is that the global working classes, with great **unwavering solidarity, would have to establish their own global economy,** based upon collective ownership, collective management of first small businesses, and gradually growing to global operations. We have to completely dissociate ourselves with any kinds of politics, ideologies and religions, and any kinds of isms that are mushrooming, that are designed to keep us entertained. We have to stay away from all kinds of false nationalism, and hypocritical patriotism, that have only brought us misery, destruction, humiliation, wars of mutual destructions. The entire global working classes form **a diverse family**, and diversity is the beauty of it. The global working classes must never fight against the rest of the family. We have to internationalize trade and intermarriages, responding to one another needs, aspirations, with conscious attempt of creating a more peaceful, and harmonized global family. There are plenty of lands, resources, and opportunities. Let us take advantage of them, in favor of mankind. **This, in a nut shell, is what Global working class economy is. It is futile to try** to change either of the two existing systems. Leave that struggle to other interested parties. Put your entire efforts on creating a decent satisfying life for yourself, by helping to create the global working class economy, and your first step would be get together with other people, to establish the kind of business you like to spend your life on, as partners. **Use private property principles** to manage, and jointly enjoy the fruits of your labor. The hell with all capitalists, the hell with all governments. You are not a beggar. Create your own business, your own life, and your future. Don't waste your life on these **ruthless bastards**. You don't need a political party, you don't need a trade union to protect you. Within your body and mind, you have the potentials to producing abundance of wealth that would last you for the rest of your life.

Chapter 8
THE NATURE OF GOVERNMENTS OF ALL POLITICAL ORIENTATIONS

According to Marxist Socialism, all governments are instruments of oppression of one dominant social class, that collectively owns the means of production, the capitalist class, against the rest of the population who survive by selling their labor power, as wage slaves. All governments survive by using different forms of oppression, and force. They use the army, the police, the educational system, the different branches of the government, the legislative, the judiciary, the executive, even the Constitution to legitimize their conduct, and guarantee their survival, and oppressive dominance. On the worldwide basis, all governments are related to one another. Their most tangible bond is based upon ownership of means of productions, that go beyond the lands in which they were born. They have developed the sophisticated legal techniques of concealing their financial economic assets, vast land ownership, and production facilities on every globally lucrative business all over the world. They don't have any feelings of loyalty and patriotism to where they were born. Their loyalty is to the ownership of the entire Planet Earth. And their unwavering love is for capital, dollars, and tons of it, and not just a few. They are literally one another's cousins, brothers and sisters, brother in laws sister in laws mother in laws father in laws, close friends, and so forth. **There is no such thing as an independent government on the face of the Earth.** Their insatiable desire to survive, and rule the entire Planet Earth would make them to have a close working relationship.

Chapter 9
MULTI-NATIONAL CORPORATIONS

The time in which small companies produced to sell to their villages, little- home towns, and cities are gone forever. In the colonial period, shoemakers made enough shoes in order to satisfy the needs of the village.

Taylors made suits, one by one, correctly measured according to bodies of the customers. In those days, the producers were in the hundreds, and they competed against one another to sell their products. The production facilities, and the ultimate buyers -users were living in the same area. Their employees, if any, did not exceed four or five individuals. **This was in the colonial era**, right around the formation of the United States, out of the colonies. The U.S. Constitution was written in that cultural intellectual agricultural mentality and background, and reflected the production capacities of that time. The country was agricultural, and starting to take the beginning steps toward industrialization. The understanding of life and production activities **were pre scientific pre-atomic age**. Nowadays, multi national corporations produce, and sell their products all over the world. Factories mass produce for millions. China, that is a component part of the Mafia of International Capitals, with the help of American, West European, Japanese, and Canadian multi billionaires are producing and selling their products to every corner of the World, most of the time in exchange for raw materials of all kinds, and global land ownership. **These corporations do not have any loyalty to any given country. The whole planet Earth is what they own, and they have the entire global masses of people as their hostages, and wage slaves in all countries.**

Chapter 10

THE MARRIAGE OF ALL GOVERNMENTS, LEFT&RIGHT, WITH MULTI-ANTIONAL CORPORATIONS FORMED THE MAFIA OF INTERNATIONAL CAPITALS

The Mafia of International Capitals owns the best and the most strategically desirable sources of raw materials and real estate properties, all mass scale production facilities, shopping centers, major apartment houses, high rises, the banks, the major communication media, the global shipping companies, everything that makes money, globally and rules the entire Planet Earth, politically and economically. They create governments, overthrow governments, designate, distribute capital and labor to the economies of all countries. They install governments, support certain individuals to rule and govern in a given country. They conduct military coups against any government that gets out of control and becomes rebellious to their global leadership. They punish, and reward certain leaders, base upon their performances, and assassinate them, if regular resolution of the issues would become undoable, and complicated. **But, at the end of the day, it is between the same family, called the Mafia of International Capitals**. At times, they pretend to have disagreements with one another, in order to confuse the public. But, deep down, they are close allies, Just as disagreements arise between professional thieves together, or the Mafia of drug trafficking, in dividing stolen goods, or disagreement over style of operation, or division of areas of influence, resulting in assassination of one another, the same thing takes place between the major players of the Mafia of International Capitals. **What bond them together are not religions, ideologies, philosophical ideas, or political orientations, it is the ownership of the Planet Earth, and complete control of all the global** natural resources, and major **production facilities on the global basis**. It gives the working classes, the impressions that they carry certain weights in the society, and they do, to a very insignificant degree, as long as they are completely subservient to the maintenance of the global leadership of the Mafia of International Capitals, but the working classes of the world are their **collective possession, and their collective slaves**. The Mafia of International Capitals manufacture literally hundreds of ideologies, political orientations, unlimited religious denominations, highlighting and magnifying their differences, and divisive features, and instilling them upon the highly impressionable minds of our children through our educational system, the media, art and culture, in order to keep us busy and entertained for ever **so that we would not have time to think and act upon the real issues, important to our own lives.**

Chapter 11

IT IS IMMORAL, & UNETHICAL FOR THE GLOBAL WORKING CLASS TO DEFEND, AND SACRIFICE THEIR OWN LIVES, AND THE LIVES OF THEIR SONS, AND DAUGHTERS FOR ANY GOVERNMENT, OR COUNTRY AGAINST OTHERS

When the Mafia of International Capitals experiences problems among its component parts, and they want to put up a challenge, and a fight, they recruit soldiers from working classes sons and daughters. They convince the working class of one country to mobilize and destroy the **"the enemy working classes of other nations"**. This is how they sacrifice the youth of all nations to settle account with their own rivals. The working class of one nation has nothing against the working classes of other nations. The issues are among the ruling classes of various countries themselves, and not among the global working classes. **The people of Afghanistan, Iraq, and Libya had nothing to do with 9/11**. Yet millions of people in these countries were destroyed, their economies ruined, thousands of American youth gave their lives, some forty thousand American soldiers came back from the war with their arms, legs, amputated, **coming back to their High school sweet hearts**, the dreams of their lives, in these deplorable conditions of human trauma. Why ? in order to remove the last vestiges of the Soviet Union influence, Iraq, Libya, and Syria, some lunatics among the Mafia of International Capitals tried to reshuffle the entire Middle East. The majority of the Arab students, who are accused of having master-minded and executed the crime of the New York twin Towers, came from Saudi Arabia, the close ally of United Sates in Middle East, and not from Afghanistan, Iraq. and Libya. Why didn't George W. Bush invade Saudi Arabia? How could nineteen Arab students who had allegedly taken some basic f light lessens in United States, as hobbies, as f light adventurous beginners, and students, the sons of Saudi multi billionaires master - mind the crashing of the two American Airline planes into the New York Towers, with such a degree of accuracy. Navigating long distance commercial f lights is not like driving a car, where someone sits behind the wheel, manipulating the movements of the car with great precision. Navigating an airplane is pre -arranged through a very sophisticated electronic system, that requires years of theoretical and practical training and expertise, and not by the pilots manually steering the wheel, as they would in a car. The Arab students did not have electronic sophistication, to navigate, regulate, and maneuver the movement of the planes. **The movements of these two planes were done outside of the plane, very much like controlling drones, electronically, from unknown places, somebody sitting behind a desk,** working on a computer, in comfortable environment. **The pilots, the passengers, and the Arab students, and the more than three thousand, completely innocent people, within the Twin Towers had no knowledge as to what was going to happen, and what their destiny would be.** Somebody hacked the movement of the planes, paralyzing, and out -smarting the planes' electronic system, completely confusing the pilots, putting them in relaxed state of mind, believing that everything was quite normal, up to the time they faced their fatal destiny They were all murdered, in cold blooded fashion, as complete victims. **The actual murderers of this crime had nothing to do with Afghanistan, Iraq, and even Ben Ladin, and his people.** The actual murderers of this horrendous crime of the twenty first century were the agents of the Mafia of International Capitals, who have not been brought to justice. The Obama's administration claim that they captured Ben Ladin in a dinky little apartment, with several of his "wives", and some porno movies are as true as the Arab students master- minding the 9/11. **Three very courageous American women, working for the C.I. A, the State Department, and Pentagon expose these lies to a great extent. They are Barbara Honegger, April Gallop, and Susan Lindauer.** Get on YouTube, and find out who these truly patriotic American people are. **April Gallop was working for Pentagon,** on the day a third plane is claimed to have crashed into Pentagon building, destroying some parts of the building and causing death among some employees. **She says she noticed some bombs going off in the building,** as soon as she struck a key on her computer's keyboard. She testified that she did not see any parts of a plane, or broken plane seats, or various parts of human bodies, even though, there are **numerous surveillance cameras,** installed all over the Pentagon's building. **What happened to literally thousands of pictures, taken by the surveillance cameras? She saw some parts of a drone.** The question is: if there was a passenger plane, with some passengers missing, and it did not crash into Pentagon building, then, **what happened to that plane and the passengers?** It may have been redirected and completely destroyed, someplace else, in order to dispose of the evidence. **The hole created on the side of the Pentagon building was a round one, and was much smaller than the width of the plane, and had been created by a bomb. She had observed some parts of a drone, and not some parts of any plane. she testified that all the alarm system, the warning system, and the special room where all decisions are made were completely turned off.**

She added that several bombs went off at different times, and that there was no evidence of any plane crashing into the pentagon building. There were no eye witnesses, in the vicinity, testifying that they had seen a plane crashing into the pentagon building either. Some testified that they had seen a plane nearing the building and MANEUVERING itself away from the building. Once again, if the American f light 757 did not crash into the building, as there are no evidences that it did, then what happened to the plane, the passengers, the remaining four Arab students, and the crew? Everything had been pre arranged by the Mafia of International Capitals. Both, George W. Bush, and Hussein Obama must be brought to justice, one for planning the 9/11, and the other for covering it up. This was a crime and insult against humanity, and not against the American people alone. They showed no mercy or compassion for the victimized Americans, Afghanis, and Iraqis. This is a concrete lessen of why the global working classes of all nations should stay together against the Mafia of International Capitals, and never defend any given government, and sacrifice their lives against, or for others, for the simple reason that all governments are an integrated family. *The best government is a dead government.* The 9/11 had nothing to do with the Arabs, the Middle East, or Ben Ladin. It was an act by the Mafia of International Capitals. Middle East will not normalize for the next 50 years, even if the entire Isis Military operations are destroyed. There will be half a dozen new radicalized organizations emerging, one by one, because their countries, their people, their honor, and dignity were destroyed by a pack of lies. Recently appeared an article on internet, saying that the Saudis admitted that they "misled their American friends" on Islamic radicalization of Saudi people. But, they said they did it to counter the socialism of JAMAL Abdel Nasser of EGYPT in the early fifties, and against Soviet Union invasion of Afghanistan, in the early eighties, and the rise of an ex- tremist Shiite government in Iran, and its influence in LEBANON, Syria, and among the PALESTINIANS in GAZA Strip.

All of these were done with the Blessing and cooperation, and planning by the U.S. Government. What part of this is "MISLEADING to the U.S. government? Recently, the U.S. Congress passed a law, enabling the survivors of 9/11 to sue Saudi Arabia for radicalizing their students to commit the crime of 9/11. This would be a legitimate demand if it were true that the Saudi students committed this crime. But, the crashing of these American Airline f lights into the twin Towers was pre -arranged electronically by outside forces, controlling the planes as if they were drones. If this is true, then the issue is the other way around, the Saudi students' survivors, and all the other American victims survivors must sue the Mafia of International Capitals, and the U.S. government, and George W. Bush for having planned this horrendous crime of the century. They should also sue President Hussein Obama for hiding this pregnancy for eight years. The gravity of this crime is such that it would go beyond the American land, it is an event against the entire humanity. It is unfortunate that the United Nations is being manipulated by the big powers, as another source of all fabrications and cover-ups. The concept of *SOVEREIGNTY* has become a vicious tool in the hands of the Mafia of International Capitals, to commit whatever crimes they wish to commit against humanity, and if the citizens of other nations object to them, they would holler that they are: *INTERFERING* in the internal affairs of other countries. It is the inviolable right of any person, living on this Planet Earth, to comment and criticize any event, happening anywhere in the world. The Mafia of International Capitals interfere in the affairs of all nations globally. while denying the same right to private individuals of all nations. There is no such thing as ½ pregnancy. we are either pregnant, or not pregnant. if we are, we have to accept the responsibility of the child being born. If we are supposed to interfere, **we all interfere**, and if we are not supposed to interfere, then we all should stay away from one another's affairs. **We are not living in caves by ourselves,** we are living as members of a community, called the "planet Earth". Gone are the times when each nation had a gate with heavy bolt locks, and soldiers around the borders, committing unimaginable crimes against their own citizens, and nobody in other nations knew that these things were happening. Historians would record them with fear and hesitations, and many decades or even centuries afterwards, people would discover of the atrocities. Now, in the age of internet, and social media, all events are known **SIMULTANEOUSLY** by all mankind. we can't have two sets of laws, one for those who could interfere, and one for those who are forbidden, and must stay away. we all have the right to interfere. we all have to know what is going on and be a part of the decision making. I don't give a damn if it is against the bible, against the Constitutions of all nations. the hell with all of them. If I was not a part of writing up the constitutions, then, I will not honor them. let the forefathers who wrote them, emerge from their graves. and honor what they wrote. they had no right to write a document for future generations. If a document written previously must be honored for ever, and for all times. why did the American forefathers, throw the British Constitution in the garbage, and rewrote a new one? Were they not interfering with affairs of the British empire who completely paid for the colonization of north America, by the sweat, and money of the British people, and the Black slaves that made it possible? why did the British dishonor the constitutions of the native Americans, and **MERCILESSLY Eliminated** a good portion of the native Americans in order to colonize more than half of the continent? constitutions are as good as the paper they are written on. completely worthless, if they were written to legitimize and institutionalize the con- quest of a land and its people. No constitutions, no documents of any kinds, including the Bible, the Old Testament, and the New should be honored, if they are not completely at the service of mankind. People them- selves are the living constitutions. We do not need a piece of paper to tell us how we should conduct our lives. All constitutions were written in order to legitimize a system of slavery, and the global working classes should completely reject them altogether. What happened to the soviet constitu- tion, one of the most completed form of **GUAR ANTEEING** every human right? A constitution must be ref lective of the actual situations, and the lives of the ordinary people in a society, and not like a bible that people could swear to when they are in trouble. Their Constitution said the working people were managing the economy. It was a great lie. It was the self assigned bureaucrats, who said the working class was too busy, managing its own affairs, and that they needed a representative, to rule their affairs. So, the Soviet Government took charge with the working class remaining as wage slaves, just as they had been in the capitalist economies. A paper version of the Soviet Constitution remains in libraries for future generation to study. Planet earth and humanity are the real constitutions. If constitutions legitimize indecency, injustice, all kinds of suffering, then, we should burn all their copies, and put them in a trash can, so not even the future generation would know what kind of deceptions they were promoting. As long as we maintain two set of books, one for the IRS, and one for ourselves, as long as we are engaged in fraud, deception, manipulation, double standard., there is no hope of having peace on global basis. *Susan Lindauer,* an American citizen, had been working for CIA., as what is called a "Special Asset", during the George W. Bush administration. She represented the U. S. Government to maintain a close contact between U.S. and the Iraqi, and Libyan governments. She carried out all the back and forth important decisions, involving highly sensitive issues, personally between George W Bush, Saddam Hussein, and Muammar Qaddafi of Libya. She TESTIFIES that almost a year before the actual 9/11 event, and the alleged attack on the Pentagon building, the U.S. Government was sending constant messages to Saddam Hussein Government, through **Susan Lindauer**. The content of the message was that The United States Government had unconfirmed rumors that some unknown people from the Middle East were trying to hijack a plane and crash it into the New York Twin Towers, and the United states believes that Saddam Hussein had that information. That message was taken from George W. Bush to Saddam Hussein many times by Susan Lindauer; and each time, Saddam Hussein would deny they had such information, and that Saddam Government was inviting the FBI. to go to Baghdad, to personally investi- gate the claims, and that the Iraqi Government would cooperate with Washington fully to pursue the issue to their satisfaction. Washington kept insisting that the Iraqi Government had the information, and that they were LYING, refusing to provide correct information. As you see, George W. Bush was spreading the news that this was going to happen, preparing the public for this eventuality, while the United States Government was making plans to do it itself. Ms. **Lindauer testifies that a month before the the 9/11. a white van had been going to the Towers late in the evening, after the cleaning crews had left, for several weeks to wire up the buildings for bombs to go off, inside.** After 9/11, she quoted a high Iraqi official as saying: "what kind of people are you, who would murder more than a few thousand of your own citizens, in order to have a pretext to invade other countries." **This lady was a U.S. Government** high official "Special Asset agent", and not an ordinary individual, out of U.S. Government. **George W. Bush put her in jail for several years, based on the laws of "Patriot Acts, one and two**, passed by the U.S. Government, curtailing the American civil liberties of anyone considered a **"security risk", without due process of law, claiming that "she was crazy, with a mental problem".** She categorically arrives at the conclusion that 9/11 was an inside job, in order to invade Afghanistan and Iraq. Another very important witness, working for the United States State Department, in different administrations, including that of George W. Bush, is the honorable, great patriotic American citizen, **Barbara Honegger.** She was an insider in the George W. Bush administration. She testifies that, a couple of years before 9/11, there was an ongoing conversation in the State Department that talked about "the new American century manifesto", claiming that very soon something very serious would happen that would enable the United States to invade the Afghanistan and Iraq, **to take control of the oil wells and reshuffle the entire Middle East. She says they were indirectly talking about 9/11 event in advance.** She also testified that no plane had crashed into the Pentagon building, that the damages made to the building **were as the result of several bombs going off at the Pentagon building. This coincides with April Gallop's testimony,** a Pentagon employee, that there was no evidence of any plane crashing into the Pentagon building. The question is what happened to the American Airline, f light 757, with the passengers, the crew, and the four remaining Arab students? **Susan Lindauer maintains that many of the Arab students who allegedly crashed the American Air lines into the Twin Towers have been spotted alive by their relatives, in different places, and countries, ironically after the 9/11.** DOESN'T the U.S. Government have the responsibility to investigate and find out where these alleged perpetrators of these hideous crimes are? **I think that they know where they are, and they even protect them to remain unknown, and not discovered. The real LEGACY of George W. Bush, and Barak Hussein Obama is that George planned the 9/11, and Obama covered it up,** and the people who carried out the mission are at large. All mankind should raise a billion dollar, as an award to any one who has the information, to solve the horrendous MYSTERY, an insult to all humanity. **I encourage everybody to go to YouTube and click on these great American heroes, April** Gallop, **Susan Lindauer, and Barbara Honegger,** and listen to their testimonies in details, and decide for yourself. It is the responsibility of every decent American, and non-American cooperate to solve this issue. **On October 08/ 2016,** around 5: PM, I was listening to a talk show, called "870 AM, the Answer", on a program, called" The **champions of Justice". Mr. Richard De vista,** an attorney on the **Water Gate,** and the person who had interviewed, both, Condoleeza Rice, and also President, George W. Bush, regarding the 9/11 incidence, and the attack on the Pentagon Building, was himself being interviewed, as a quest of the program, on the same issue. He stated and confirmed the conversation that he had with President Bush, and his foreign policy advisor, Condoleeza Rice, before she became the Secretary of State. Attorney Richard De vista asked the President if he had received any warnings from F. B. I, and the C. I. A., prior to the 9/11 incidence. Both admitted that sometime **in August, a few weeks before the 9/11,** they had received a briefing from F. B. I., that a few Arab students, with one or two of them having received f light trainings, on a one engine plane, were planning to high jack a plane and crash it into the New York Twin Towers. **Attorney Richard De vista asked President Bush: "why didn't you try to prevent that". President**

George W. Bush responded that he never thought a bunch of Arab students, with a few of them having learned to f ly a one engine plane, would be capable to do such monumental task, without having the scientific sophistication to navigate a passenger commercial plane. This is precisely what I have been arguing, that navigating a commercial plane, with very sophisticated electronic system, requires years of scientific trainings, and experience, both theoretical and practical, and that a person with a limited hours of beginners' flight trainings is not experienced enough to crash a commercial plane into any focused entity, with great precision, as was the case. It is my strong belief that these two planes were crashed into the Towers, being electronically guided from outside, very much like guiding the movements of a drone, and that all the people in the plane, including the crew, and the Arab students, were completely unaware, being victims of a conspiracy to commit a horrendous act. Bin Laden, and Al Qaeda, did not plan this, but upon receiving this information, they were very happy of being erroneously considered super heroes, completely invincible, allegedly bringing the only super power, militarily, and economically, to its knees. This is completely in line with the boastful mentality of the Arabs, taking credit for something that they never did. *The end.*

We would be less than honest, if we write in order to please others

ISBN: 1974371107
ISBN 13: 9781974371105